READY-TO-USE
CHEMISTRY ACTIVITIES
FOR GRADES 5-12

D1314414

Mark J. Handwerker, Ph.D.

**THE CENTER FOR APPLIED
RESEARCH IN EDUCATION**
West Nyack, New York 10994

Library of Congress Cataloging-in-Publication Data

Handwerker, Mark J.
    Ready-to-use chemistry activities for grades 5–12 / Mark J.
Handwerker.
        p.   cm — (Secondary science curriculum activities library)
        ISBN 0-87628-438-1
        1. Chemistry—Study and teaching (Middle school)   2. Chemistry—
Study and teaching (Secondary)   I.   Title.   II.   Series.
QD40.H325   1998
540'.7'12—dc21

98-35566
CIP

Printed in the United States of America

10   9   8   7   6   5   4   3   2   1

ISBN 0-87628-438-1

ATTENTION: CORPORATIONS AND SCHOOLS

The Center for Applied Research in Education books are available at quantity discounts with bulk purchase for educational, business, or sales promotional use. For information, please write to: Prentice Hall Direct Special Sales, 240 Frisch Court, Paramus, NJ 07652. Please supply: title of book, ISBN number, quantity, how the book will be used, date needed.

**THE CENTER FOR APPLIED RESEARCH
IN EDUCATION**
West Nyack, NY 10994

On the World Wide Web at http://www.phdirect.com

Prentice-Hall International (UK) Limited, *London*
Prentice-Hall of Australia Pty. Limited, *Sydney*
Prentice-Hall Canada, Inc., *Toronto*
Prentice-Hall Hispanoamericana, S.A., *Mexico*
Prentice-Hall of India Private Limited, *New Delhi*
Prentice-Hall of Japan, Inc., *Tokyo*
Simon & Schuster Asia Pte. Ltd., *Singapore*
Editora Prentice Hall do Brasil, Ltda., *Rio de Janeiro*

About This Resource

Ready-to-Use Chemistry Activities for Grades 5–12 is designed to help you teach basic science concepts to your students while building their appreciation and understanding of the work of generations of curious scientists. Although The Scientific Method remains the most successful strategy for acquiring and advancing the store of human knowledge, science is—for all its accomplishments—still merely a human endeavor. While the benefits of science are apparent in our everyday lives, its resulting technology could endanger the survival of the species if it is carelessly applied. It is therefore essential that our students be made aware of the nature of scientific inquiry with all its strengths and limitations.

A primary goal of science instructors should be to make their students "science literate." After completing a course of study in any one of the many scientific disciplines, students should be able to:

1. appreciate the role played by observation and experimentation in establishing scientific theories and laws,

2. understand cause-and-effect relationships,

3. base their opinions on fact and observable evidence—not superstitions or prejudice, and

4. be willing to change their opinions based on newly acquired evidence.

Scientific theories come and go as new observations are made. During the course of instruction, teachers should emphasize the "process" of science as well as the relevance of pertinent facts.

This volume of science activities was designed to accomplish all of the above, keeping in mind the everyday challenges faced by classroom instructors.

On Your Mark!

Begin by stimulating students' gray matter with basic scientific concepts through brainstorming and open discussion.

Get Set!

Kindle interest by making concepts real through demonstration and/or descriptive analogy.

Go!

Cement concepts into concrete form with exciting hands-on experience.

Each of the 15 teaching units in this volume of *Ready-to-Use Chemistry Activities for Grades 5–12* contains *four* 40–50 minute lessons and follows the same instructional sequence so that your students will always know what is expected of them. Each unit comes complete with the following:

- a **Teacher's Classwork Agenda for the Week** and **Content Notes for Lecture and Discussion,**

- a student **Fact Sheet** with **Homework Directions** on the back,

- four 40–50 minute **Lesson Plans,** each followed by its own **Journal Sheet** to facilitate student notetaking, and

- an end-of-the-unit **Review Quiz.**

Each unit has been tested for success in the classroom and is ready for use with minimal preparation on your part. Simply make as many copies of the Fact Sheet with Homework Directions, Journal Sheets, and Review Quizzes as you need for your class. Also, complete answer keys for the homework assignments and unit quiz are provided at the end of the Teacher's Classwork Agenda for the Unit.

Mark J. Handwerker

ABOUT THE AUTHOR

Mark J. Handwerker (B.S., C.C.N.Y., Ph.D. in Biology, U.C.I.) has taught secondary school science for 15 years in the Los Angeles and Temecula Valley Unified School Districts. As a mentor and instructional support teacher, he has trained scores of new teachers in the "art" of teaching science. He is the author/editor of articles in a number of scientific fields and the coauthor of an earth science textbook (Harcourt Brace Jovanovich, *Earth Science*) currently in use.

Dr. Handwerker teaches his students that the best way to learn basic scientific principles is to become familiar with the men and women who first conceived them. His classroom demonstrations are modeled on those used by the most innovative scientists of the past. He believes that a familiarity with the history of science, and an understanding of the ideas and methods used by the world's most curious people, are the keys to comprehending revolutions in modern technology and human thought.

Suggestions for Using These Science Teaching Units

The following are practical suggestions for using the 15 teaching units in this resource to maximize your students' performance.

Fact Sheet

At the start of each unit, give every student a copy of the **Fact Sheet** for that unit with the **Homework Directions** printed on the back. The Fact Sheet introduces content vocabulary and concepts relevant to the unit. You can check students' homework on a daily basis or require them to manage their own "homework time" by turning in all assignments at the end of the unit. Most of the homework assignments can be completed on a single sheet of standard-sized (8½" × 11") looseleaf paper. Urge students to take pride in their accomplishments and do their most legible work at all times.

Journal Sheet

At the start of each lesson, give every student a copy of the appropriate **Journal Sheet** which they will use to record lecture notes, discussion highlights, and laboratory activity data. Make transparencies of Journal Sheets for use on an overhead projector. In this way, you can model neat, legible, notetaking skills.

Current Events

Since science does not take place in a vacuum (and also because it is required by most State Departments of Education), make **Current Events** a regular part of your program. Refer to the brief discussion on "Using Current Events to Integrate Science Instruction Across Content Areas" in the Appendix.

Review Quiz

Remind students to study their Fact and Journal Sheets to prepare for the end-of-the-unit **Review Quiz.** The Review Quiz is a 15-minute review and application of unit vocabulary and scientific principles.

Grading

After completing and collectively grading the end-of-the-unit Review Quiz in class, have students total their own points and give themselves a grade for that unit. For simplicity's sake, point values can be awarded as follows: a neatly completed set of Journal Sheets earns 40 points; a neatly completed Homework Assignment earns 20 points; a neatly completed Current Event earns 10 points; and, a perfect score on the Review Quiz earns 30 points. Students should record their scores and letter grades on their individual copies of the **Grade Roster** provided in the Appendix. Letter grades for each unit can be earned according to the following point totals: A ≥ 90, B ≥ 80, C ≥ 70, D ≥ 60, F < 60. On the reverse side of the Grade Roster, students will find instructions for calculating their "grade point average" or "GPA." If they keep track of their progress, they will never have to ask "How am I doing in this class?" They will know!

Unit Packets

At the end of every unit, have students staple their work into a neat "unit packet" that includes their Review Quiz, Homework, Journal Sheet, Current Event, and Fact Sheet. Collect and examine each student's packet, making comments as necessary. Check to see that students have awarded themselves the points and grades they have earned. You can enter individual grades into your record book or grading software before returning all packets to students the following week.

You will find that holding students accountable for compiling their own work at the end of each unit instills a sense of responsibility and accomplishment. Instruct students to show their packets and Grade Roster to their parents on a regular basis.

Fine Tuning

This volume of *Ready-to-Use Chemistry Activities for Grades 5–12* was created so that teachers would not have to "reinvent the wheel" every week to come up with lessons that work. Instructors are advised and encouraged to fine tune activities to their own personal teaching style in order to satisfy the needs of individual students. You are encouraged to supplement lessons with your district's adopted textbook and any relevant audiovisual materials and computer software. Use any and all facilities at your disposal to satisfy students' varied learning modalities (visual, auditory, kinesthetic, and so forth).

CONTENTS

CH1   THE PROPERTIES AND PHASES OF MATTER / 1

Teacher's Classwork Agenda and Content Notes

Classwork Agenda for the Week . . . Content Notes for Lecture
and Discussion . . . Answers to the End-of-the-Week Review Quiz

Fact Sheet with Homework Directions

Lesson #1
Students will discuss the common properties of matter and how heat energy causes matter to change phase.
Journal Sheet #1

Lesson #2
Students will explain how early thermometers were invented and show how to calibrate a thermometer by measuring the temperature at which water vaporizes.
Journal Sheet #2

Lesson #3
Students will graph the temperature of water as it changes from a liquid to a solid.
Journal Sheet #3

Lesson #4
Students will explain and demonstrate that evaporation is a "cooling" process.
Journal Sheet #4

CH1 Review Quiz

CH2   MIXTURES / 15

Teacher's Classwork Agenda and Content Notes

Classwork Agenda for the Week . . . Content Notes for Lecture
and Discussion . . . Answers to the End-of-the-Week Review Quiz

Fact Sheet with Homework Directions

Lesson #1
Students will discuss the nature of mixtures and begin an experiment that shows that air is a mixture of different gases.
Journal Sheet #1

Lesson #2
Students will identify the various components of a solid mixture.
Journal Sheet #2

Lesson #3
Students will separate the components of a solution by distillation.

Journal Sheet #3

Lesson #4
Students will demonstrate that different substances dissolve at different rates and that temperature affects the solubility of substances.
Journal Sheet #4

CH2 Review Quiz

CH3 CRYSTALS / 29

Teacher's Classwork Agenda and Content Notes

Classwork Agenda for the Week . . . Content Notes for Lecture
and Discussion . . . Answers to the End-of-the-Week Review Quiz

Fact Sheet with Homework Directions

Lesson #1
Students will show how to grow crystals in class and at home.
Journal Sheet #1

Lesson #2
Students will build toothpick and construction paper models of crystals.
Journal Sheet #2

Lesson #3
Students will make accurate drawings of the salt crystals they observe under the microscope or magnifying glass.
Journal Sheet #3

Lesson #4
Students will make accurate drawings of the crystals found in common rocks.
Journal Sheet #4

CH3 Review Quiz

CH4 HEAT AND ENERGY TRANSFER / 43

Teacher's Classwork Agenda and Content Notes

Classwork Agenda for the Week . . . Content Notes for Lecture
and Discussion . . . Answers to the End-of-the-Week Review Quiz

Fact Sheet with Homework Directions

Lesson #1
Students will be able to explain the difference between heat and temperature and show how heat energy is conducted in a solid.
Journal Sheet #1

Lesson #2
Students will show how heat energy is convected through a liquid or gas.
Journal Sheet #2

Lesson #3
Students will show how heat radiation can be concentrated at a focal point.
Journal Sheet #3

Lesson #4
Students will measure the number of calories in a soda cracker.
Journal Sheet #4

CH4 Review Quiz

CH5 PHYSICAL AND CHEMICAL CHANGE / 57

Teacher's Classwork Agenda and Content Notes

Classwork Agenda for the Week . . . Content Notes for Lecture
and Discussion . . . Answers to the End-of-the-Week Review Quiz

Fact Sheet with Homework Directions

Lesson #1
Students will be able to explain the difference between a physical and a chemical change.
Journal Sheet #1

Lesson #2
Students will separate two simple mixtures by physical means.
Journal Sheet #2

Lesson #3
Students will perform two chemical reactions.
Journal Sheet #3

Lesson #4
Students will be able to distinguish between exothermic and endothermic chemical reactions.
Journal Sheet #4

CH5 Review Quiz

CH6 ELEMENTS, MOLECULES, AND COMPOUNDS / 71

Teacher's Classwork Agenda and Content Notes

Classwork Agenda for the Week . . . Content Notes for Lecture
and Discussion . . . Answers to the End-of-the-Week Review Quiz

Fact Sheet with Homework Directions

Lesson #1
Students will use chemical symbols and formulas to illustrate the difference between elements, molecules, and compounds.
Journal Sheet #1

Lesson #2
Students will build models to visualize the difference between elements, molecules, and compounds.
Journal Sheet #2

Lesson #3
Students will demonstrate that water is a compound.
Journal Sheet #3

Lesson #4

Students will write and balance chemical equations to show how matter is conserved in a chemical change.

Journal Sheet #4

CH6 Review Quiz

CH7 ATOMIC STRUCTURE / 85

Teacher's Classwork Agenda and Content Notes

Classwork Agenda for the Week . . . Content Notes for Lecture
and Discussion . . . Answers to the End-of-the-Week Review Quiz

Fact Sheet with Homework Directions

Lesson #1

Students will draw diagrams of atoms to show how models of atoms have evolved.

Journal Sheet #1

Lesson #2

Students will construct models of Bohr atoms.

Journal Sheet #2

Lesson #3

Students will diagram how atoms can be transformed into ions.

Journal Sheet #3

Lesson #4

Students will diagram how atoms can form ionic or covalent bonds with other atoms.

Journal Sheet #4

CH7 Review Quiz

CH8 THE PERIODIC TABLE OF THE ELEMENTS / 99

Teacher's Classwork Agenda and Content Notes

Classwork Agenda for the Week . . . Content Notes for Lecture
and Discussion . . . Answers to the End-of-the-Week Review Quiz

Fact Sheet with Homework Directions

Lesson #1

Students will be able to explain how elements are arranged on The Periodic Table.

Journal Sheet #1

Lesson #2

Students will be able to compare and contrast the physical and chemical properties of families on The Periodic Table.

Journal Sheet #2

Lesson #3

Students will prepare a group report on the elements of a chemical family.

Journal Sheet #3

Lesson #4
Students will present a group report on the elements of a chemical family.
Journal Sheet #4

CH8 Review Quiz

CH9   TYPES OF CHEMICAL REACTIONS / 113

Teacher's Classwork Agenda and Content Notes

Classwork Agenda for the Week . . . Content Notes for Lecture
and Discussion . . . Answers to the End-of-the-Week Review Quiz

Fact Sheet with Homework Directions

Lesson #1
Students will distinguish between a synthesis and a decomposition reaction.
Journal Sheet #1

Lesson #2
Students will perform a decomposition reaction that liberates explosive oxygen gas.
Journal Sheet #2

Lesson #3
Students will capture an explosive gas as the product of a single displacement reaction.
Journal Sheet #3

Lesson #4
Students will form a precipitate as the product of a double displacement reaction.
Journal Sheet #4

CH9 Review Quiz

CH10   ACID-BASE REACTIONS / 127

Teacher's Classwork Agenda and Content Notes

Classwork Agenda for the Week . . . Content Notes for Lecture
and Discussion . . . Answers to the End-of-the-Week Review Quiz

Fact Sheet with Homework Directions

Lesson #1
Students will list the physical and chemical properties of acids and bases.
Journal Sheet #1

Lesson #2
Students will use litmus paper to measure the pH of acids and bases.
Journal Sheet #2

Lesson #3
Students will demonstrate that acids and bases form ionic solutions.
Journal Sheet #3

Lesson #4
Students will neutralize a base by titration with an acid.
Journal Sheet #4

CH10 Review Quiz

CH15 NUCLEAR REACTIONS AND RADIOACTIVITY / 197

Teacher's Classwork Agenda and Content Notes

Classwork Agenda for the Week . . . Content Notes for Lecture
and Discussion . . . Answers to the End-of-the-Week Review Quiz

Fact Sheet with Homework Directions

Lesson #1
Students will draw models of the structure of atomic nuclei and explain what is meant by the term "isotope."
Journal Sheet #1

Lesson #2
Students will describe the different types of radiation and write nuclear equations.
Journal Sheet #2

Lesson #3
Students will explain how the "half-life" of atomic isotopes is used to find the age of ancient objects.
Journal Sheet #3

Lesson #4
Students will build models of atomic nuclei and illustrate the difference between fission and fusion.
Journal Sheet #4

CH15 Review Quiz

APPENDIX / 211

Aristotle
Svante August Arrhenius
Amadeo Avogadro
Leo Hendrick Baekeland
Henri Becquerel
Claude Bernard
Jöns Jakob Berzelius
Joseph Black
Niels Bohr
Ludwig Eduard Boltzman
Jean-Baptiste Boussingault
Robert Boyle
William & Lawrence Bragg
Robert Brown
Robert William Bunsen
Wallace Hume Carothers
Henry Cavendish
Anders Celsius
James Chadwick
Hilaire B. Chardonnet
Michel-Eugéne Chevruel
John Douglas Cockcroft
Francis H. C. Crick
Marie Curie
Pierre Curie
Louis Daguerre
John Dalton
Abraham Darby
Charles Robert Darwin
Humphrey Davy
Democritus
René Descartes
Edwin Laurentine Drake
Paul Ehrlich
Albert Einstein
Empedocles
Gabriel Daniel Fahrenheit

Michel Faraday
Enrico Fermi
Elizabeth Fulhame
Galen
Galileo Galilei
Pierre Gassendi
Hans Geiger
Donald Arthur Glaser
Johann Rudolf Glauber
William Harvey
René-Just Haüy
Werner Karl Heisenberg
Germain Hess
Hippocrates
Robert Hooke
Ernst F. Hoppe-Seyler
John Wesley Hyatt
Jan Ingenhousz
William Thomson Kelvin
Irving Langmuir
Pierre Simon Laplace
August Laurent
Antoine Laurent Lavoisier
Anton von Leeuwenhoek
Gottfried Wilhelm Leibniz
Justus von Liebig
Carolus Linnaeus
Andreas S. Marggraff
James Clerk Maxwell
Elmer Verner McCollum
Ferdinando Dé Medici
Dmitri I. Mendeleev
Lothar Meyer
Stanley Lloyd Miller
Henry G. J. Moseley
John A. R. Newlands
Sir Isaac Newton

J. Robert Oppenheimer
Alexander Parkes
Louis Pasteur
Linus Carl Pauling
Max Karl Ernst Planck
Joseph Priestley
Joseph Louis Proust
William Prout
Theodore W. Richards
Gilles P. de Roberval
Wilhelm Konrad Röntgen
Benjamin T. Rumford
Ernst Rutherford
Julius von Sachs
Horace & Nicholas de
  Saussure
Andreas F. W. Schimper
Matthias Jakob Schleiden
Jean Senebier
Soren P. L. Sorensen
Georg Ernst Stahl
Jean Servais Stas
Kekúle von Stradonitz
Thales
Joseph John Thomson
Harold Clayton Urey
Andreas Vesalius
Johannes van der Waals
Ernest Walton
James Dewey Watson
James Watt
Heinrich Otto Wieland
Richard Willstätter
Charles T. R. Wilson
Friedrich Wöhler
William Hyde Wollaston

THE PROPERTIES AND PHASES OF MATTER

TEACHER'S CLASSWORK AGENDA AND CONTENT NOTES

Classwork Agenda for the Week

1. Students will discuss the common properties of matter and how heat energy causes matter to change phase.
2. Students will explain how early thermometers were invented and show how to calibrate a thermometer by measuring the temperature at which water vaporizes.
3. Students will graph the temperature of water as it changes from a liquid to a solid.
4. Students will explain and demonstrate that evaporation is a "cooling" process.

Content Notes for Lecture and Discussion

Chemistry is the study of the properties and interactions of different forms of matter. It arose out of people's natural curiosity about the way substances change when mixed and mingled together. Chemistry is an **empirical science** which has led to the development of a variety of technologies that have improved humanity's quality of life since ancient times. The early Egyptians and Mesopotamians practiced chemistry as early as 3,000 B.C. The Egyptians produced **bronze alloy** by heating and mixing ores of **copper** and **tin**, their practice of **metallurgy** leading to the fabrication of more practical tools, weapons and armour, and ornaments to adorn their bodies and please their eyes.

The science of chemistry advanced more quickly with the invention of devices able to weigh the samples to be mixed and studied. Ancient Mesopotamians invented the **balance** out of necessity around 4,000 B.C. to make trade among merchants more exacting and equitable. The Mesopotamian balance was a wooden beam set atop a central pivot that allowed merchants to match and compare masses of different substances. In 1,500 B.C., the Egyptians improved the design by suspending a plumb bob from the pivot and running the cords attached to the weighing pans through a hollow beam as shown in Illustration A. This combination of improvements

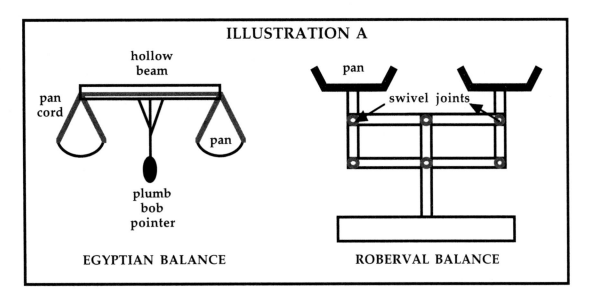

ILLUSTRATION A

hollow beam

pan cord

pan

plumb bob pointer

EGYPTIAN BALANCE

pan

swivel joints

ROBERVAL BALANCE

allowed for more accurate measurements of precious metals such as silver and gold. In 1669, French mathematician **Gilles Personne de Roberval** (b. 1602; d. 1675) invented a balance that allowed the weighing pans to balance perfectly no matter where the known masses and samples were placed in the measuring pans. This design further improved the accuracy of measurement. Modern measuring devices can mass samples to within millionths of a gram.

The Ancient Greeks developed the first **atomic theories of matter** which are so critical to our understanding of chemistry today, and are covered in later units of this volume. But it was obvious even to early metallurgists that **heat** played an integral part in the transformation of matter from one form to another. It was apparent that the addition of heat to a solid turned that solid into a liquid and that further heating could transform the liquid into a vapor. A device for measuring the amount of energy in a substance became a necessity in exploring this phenomenon. **Galileo Galilei** (b. 1564; d. 1642) designed a **thermoscope** in 1593 like the one shown in Illustration B. In 1654, **Ferdinando Dé Medici**, the fifth Grand Duke of Tuscany (b. 1610; d. 1670), invented the first **sealed thermometer**. Sealing an expandable fluid inside a glass tube eliminated interference from air pressure when measuring the amount of energy in a substance. German scientist **Gabriel Daniel Fahrenheit** (b. 1686; d. 1736) invented the first accurate thermometer in 1724 and a temperature scale to go along with it. The **centigrade scale** was introduced by Swedish astonomer **Anders Celsius** (b. 1701; d. 1744) in 1742. **William Thomson Kelvin** (b. 1824; d. 1907), who substantiated the **Second Law of Thermodynamics** by demonstrating that heat could not pass from a cold body to a warmer one,

ILLUSTRATION B

GALILEO'S THERMOSCOPE

As air in the glass tube and bulb cooled it contracted. Atmospheric pressure on the surface of the liquid in the large glass basin forced the colored liquid up the tube. The cooler the outside air, the higher the liquid rose up the tube.

gave us the **Kelvin scale**. The kelvin scale is the absolute scale of temperature, measuring the energy content of substances which contain hardly no energy at all. The temperature at which this can theoretically occur is called **absolute zero**.

In Lesson #1, students will list the common properties of matter and distinguish between the four phases of matter: **solid, liquid, vapor,** and **plasma**. They will also practice "massing" common objects on a laboratory balance.

In Lesson #2, students will compare and contrast the **Fahrenheit, Celsius,** and **Kelvin thermometers** and calibrate a thermometer by measuring the boiling point of water. They will identify **vaporization** as the "phase change" taking place in this activity.

In Lesson #3, students will graph the temperature of water as it freezes.

In Lesson #4, students will demonstrate that evaporation is a "cooling" process by comparing the rates of evaporation of water and alcohol.

ANSWERS TO THE HOMEWORK PROBLEMS

Students will discover that water expands as it freezes. The straw placed on top of the refrigerator acted as the "control" in this experiment. Point out that all experiments are comparisons.

ANSWERS TO THE END-OF-THE-WEEK REVIEW QUIZ

1. four	6. true	11. sublimation	16. solid
2. solids	7. true	12. true	17. liquid
3. liquids	8. true	13. true	18. vapor
4. true, gas, or vapor	9. true	14. physical	19. melting
5. plasma	10. true	15. chemical	20. water (i.e., melts at 0° C and boils at 100° C)

CH1 FACT SHEET

THE PROPERTIES AND PHASES OF MATTER

CLASSWORK AGENDA FOR THE WEEK

(1) List the common properties of matter and contrast its various phases.
(2) Compare temperature scales and calibrate a thermometer.
(3) Graph the temperature of water as it changes from a liquid to a solid.
(4) Demonstrate that evaporation is a cooling process.

The universe is made up of two things: **energy** and **matter**. All forms of energy have the ability to do work and there are many different kinds of energy: heat energy, light energy, electrical energy, mechanical energy, and so on. **Matter** is the material substance that makes up all the objects around us. Although matter can exist in many different forms, all forms of matter have some basic properties in common. First, all forms of matter take up space. They have **volume**. Second, all forms of matter have mass. **Mass** is the amount of matter in an object which can be measured using a tool called a **balance**. **Sir Isaac Newton** (b. 1642; d. 1727) discovered that all objects are attracted to other objects by the force of gravity. The force of attraction that one object has for another object is called **weight**. Since all matter has mass, all matter also has weight in a gravitational field.

Matter exists in four states or phases: **solid**, **liquid**, **gas**, or **plasma**. *Solids*, like rock and glass, have definite shape and definite volume. *Liquids*, like water and mercury, can change shape while retaining the same volume. *Gases*, like oxygen and carbon dioxide, change both shape and volume. *Plasma* is a highly energized gas made of electrically charged particles like electrons and protons. The trillions upon trillions of particles radiating toward earth from the sun—called the solar wind—is an example of a plasma.

Matter can change from one form or **phase** to another. Heating a solid causes the **atoms** that make up the solid to move faster. As the atoms absorb heat energy they move around more freely turning the solid into a liquid. This process is called **melting**. When the atoms of a liquid absorb heat they can leave the liquid entirely to form a **gas** or vapor. This process is called **vaporization**. Some solids, such as frozen carbon dioxide (commonly called dry ice), change directly into a gas when warmed. This process is called **sublimation**. Heating a gas may cause the tiny charged particles inside atoms to separate. *Electrons* leave their orbits around the atom's *nucleus* and the gas becomes an electrically charged **plasma**. **Condensation** is a term used to describe the cooling of a gas to form a liquid. **Freezing** describes the cooling of a liquid to form a solid. All of these changes are **physical changes**. In a physical change the shape or form of the matter may change but it remains the same kind of matter. Ice, water, and steam (i.e., solid, liquid, and vapor) are all simply different forms of water.

Homework Directions

Perform the following experiment to demonstrate the effects of extreme cold on liquid water.

1. Fill two plastic or paper straws with water.

2. Plug the ends of each straw with clay or chewed gum. Make sure there are no air bubbles left inside either of the straws. Draw the two straws lying side by side; label one straw "Straw A" and the other "Straw B." Label the drawing "BEFORE."

3. Wrap each straw in a paper towel. Place Straw A in the freezer compartment of a refrigerator. Place Straw B on a paper plate on top of the refrigerator.

4. Examine the straws 24-hours later. Draw the straws as you did the day before and label the drawing "AFTER."

5. Describe what happened to the water in each straw and draw a conclusion about the effects of extreme cold on the liquid water.

Assignment due: _____

_____   _____   ____/____/____
         Student's Signature   Parent's Signature   Date

THE PROPERTIES AND PHASES OF MATTER

Work Date: ____/____/____

LESSON OBJECTIVE

Students will discuss the common properties of matter and how heat energy causes matter to change phase.

Classroom Activities

On Your Mark!

Point out that the universe is comprised of two things: **energy** and **matter**. Define <u>energy as the ability to do work</u> and <u>matter as the material substance that makes up all objects</u>. Ask students the following question: How does one measure the amount of matter in an object? Answer: By using a **standard for mass called a gram** and a device called a **balance** to compare unknown masses to known ones. <u>One gram is the amount of matter in one cubic centimeter of water at 4° Celsius at sea level</u>. Lead a discussion in which students decide what <u>all</u> pieces of matter have in common. Have them refer to their Fact Sheet which mentions that all forms of matter have (1) **volume** and (2) **mass**. Show them three different solid objects. Ask: Can the shape and volume of these objects be changed without breaking them apart? Answer: No. They have constant shape and volume. Show them three graduated cylinders (or convenient containers) of different sizes. Pour the same amount of colored water (i.e., 100 ml) into the containers. Emphasize that the three containers contain the same volume of liquid but that the liquids have taken the shapes of their containers. Liquids have constant volume but indefinite shape. Show them the device in Illustration A. Point out that the air inside the flask has shape and volume. Rub your hands together to heat your palms with friction and wrap your hands around the flask as shown. What causes the liquid to rise up the tube? Answer: The air (i.e., a vapor) inside the flask has changed shape and volume. It has expanded and forced the liquid up the glass tube.

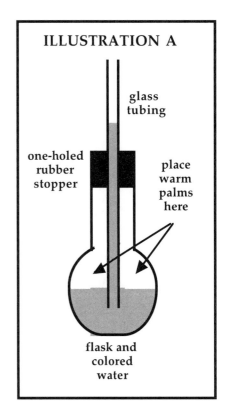

ILLUSTRATION A

glass tubing

one-holed rubber stopper

place warm palms here

flask and colored water

Get Set!

After completing Table A and taking lecture notes on Journal Sheet #1 show students how to use a balance.

Go!

Give them time to complete the simple activities in Journal Sheet #1.

Materials

flasks, one-holed rubber stopper, colored water, glass tubing, graduated cylinders, sample solid objects (rocks, coins, etc.), balance

CH1 JOURNAL SHEET #1

THE PROPERTIES AND PHASES OF MATTER

Table A			
phase	constant shape (yes or no)	constant volume (yes or no)	example
solid			
liquid			
vapor			
plasma			

DIRECTIONS

Mass Comparisons

Measure the masses of a variety of objects to practice using a balance. Make a table to record your measurements.

Volume Comparisons

(1) Mix a drop of red food coloring in water and pour about 100 ml of the colored water into a large graduated cylinder (500 ml cylinder). (2) Mix a drop of blue food coloring in water and pour about 50 ml of the water in a small graduated cylinder (100 ml cylinder). (3) Pour the water from the cylinders into two separate glasses, cups, or beakers. (4) Pour the blue water into the large beaker and the red water into the small beaker. (5) Answer the following question: (A) Did the red and blue liquids take up different amounts of space when they were transferred from one cylinder to the other? (B) Did the shape of the liquids change when they were transferred from one cylinder to the other?

FIGURE A: DOUBLE BEAM BALANCE

balance indicator

mass tray   object tray

small mass   scale accurate to 0.1 gram

large mass   scale accurate to 10 grams

THE PROPERTIES AND PHASES OF MATTER

Work Date: ____/____/____

LESSON OBJECTIVE

Students will explain how early thermometers were invented and show how to calibrate a thermometer by measuring the temperature at which water vaporizes.

Classroom Activities

On Your Mark!

Lead a brief discussion of the factors the cause matter to change form or **phase**. **Heat energy** is the primary factor leading to changes in matter. According to the **Atomic-Molecular Theory of Matter**, all material objects are made of tiny particles called atoms which are always in motion. As atoms absorb heat energy they move faster, collide with more force, and move farther apart. Draw Illustration B on the board and have students explain what happens to the atoms in a solid, liquid, and vapor as they

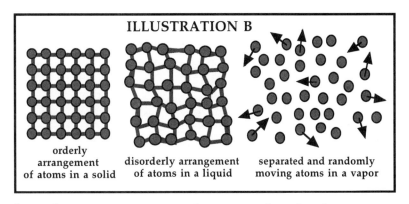

ILLUSTRATION B

orderly arrangement of atoms in a solid

disorderly arrangement of atoms in a liquid

separated and randomly moving atoms in a vapor

are heated. Explain that a **thermometer** measures the energy of motion (i.e., average kinetic energy) of trillions of atoms as they bombard the surface of the thermometer. The energy of moving atoms is transferred to the material making up the thermometer causing the liquid in the device to expand and rise up the thermometer tube.

Get Set!

Have students refer to Figure B on Journal Sheet #2. Explain that the **Fahrenheit, Celsius**, and **Kelvin** scales are simply three different "number scales" used to measure temperature. On the Fahrenheit scale water freezes at 32° F and boils at 212° F. On the Celsius scale water freezes at 0° C and boils at 100° C. On the Kelvin scale water freezes at 373K and boils at 473K. The term degrees is not used in the Kelvin scale (i.e., we say "273 kelvin"). Have students fill in the temperatures for freezing and boiling water next to each thermometer in Figure B on Journal Sheet #2. Explain that thermometers are not always perfect; so before using a thermometer, a scientist needs to determine how much it might be in error. Since water boils at 100° C at sea level, we can see if our classroom thermometers are accurate. Go over Classroom Safety Rules and General Safety Precautions when dealing with boiling water.

Go!

Have students construct the set-up shown in Figure C on Journal Sheet #2 and complete the activity as directed. Be sure they mark off "time intervals" on the "x" axis of the graph and "temperature readings" (i.e., from 20° C to 120° C) on the "y" axis.

Materials

ring stand and clamps, Ehrlenmeyer flasks, Celsius thermometers, Bunsen burners, water, heat resistant gloves, lab apron, tongs, goggles, matches

CH1 JOURNAL SHEET #2

THE PROPERTIES AND PHASES OF MATTER

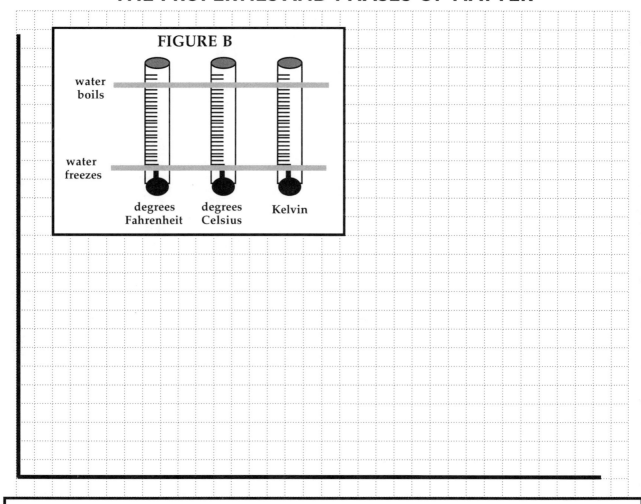

FIGURE B

water boils

water freezes

degrees Fahrenheit   degrees Celsius   Kelvin

FIGURE C

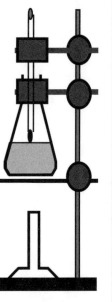

Directions: (1) Pour 50 ml of water into an Ehrlenmeyer flask. (2) Place the flask onto a ring stand and secure it with a clamp so that it cannot be toppled. (3) Secure a thermometer in a clamp and attach it to the ring stand as shown. Be sure that the tip of the thermometer barely touches the surface of the water. (4) Turn on the Bunsen burner and graph the temperature on the thermometer every 60 seconds. (5) When the water reaches a vigorous boil continue reading the thermometer for two more minutes. NOTE: Read the thermometer from a distance. Do not get your face up close to apparatus. The boiling point of pure water at sea level is 100° Celsius. If your thermometer does not read exactly 100° Celsius you must take that into account whenever you use it in an experiment. What did your thermometer read when the water boiled? What factors beside a faulty thermometer could account for one that does not read exactly 100° Celsius in boiling water?

GENERAL SAFETY PRECAUTIONS

Be sure you are familiar with the proper use of a Bunsen burner. Wear goggles and a lab apron to protect your skin and eyes from being burned by SCALDING HOT STEAM. Do not touch any part of the equipment without heat resistant gloves or tongs. Clean up when the apparatus is cool.

CH1 Lesson #3

THE PROPERTIES AND PHASES OF MATTER

Work Date: ____/____/____

LESSON OBJECTIVE

Students will graph the temperature of water as it changes from a liquid to a solid.

Classroom Activities

On Your Mark!

Discuss and define the terms **melting**, **freezing**, **vaporization** (or evaporation) and **condensation**. <u>Melting</u> describes the process of turning a solid into a liquid. <u>Freezing</u> is the opposite process. <u>Vaporization</u> is the changing of liquid to a vapor. <u>Condensation</u> is the opposite of vaporization. Have students write a sentence in which they describe having witnessed each process. Emphasize that each of these changes is a **physical** process that does not involve a changing of the substances themselves. Ice melts to form liquid water which upon further heating will vaporize to form steam. However, ice, water, and steam are <u>still</u> water! Phase change is a **physical change** and <u>not</u> a **chemical change**. In a chemical change, the substances are transformed into entirely new substances. If you have access to dry ice (frozen carbon dioxide) bring a sample to class. Explain that dry ice changes phase from a solid directly to a vapor. This process is called **sublimation**. You can also demonstrate sublimation by warming a few crystals of resublimed iodine (obtained from a laboratory supply house) in a test tube. *Avoid inhaling the purple gas that is produced. It is toxic!*

Get Set!

Draw Illustration C on the board and have students make a light pencil drawing of the graph on Journal Sheet #3. They can collect their activity data on the same graph using pen or colored pencils. Explain that when melting begins heat is required to change the phase of the substance and the energy absorbed by the atoms does not register as a rise in temperature. The temperature graph goes flat for several minutes. The same phenomenon occurs during vaporization.

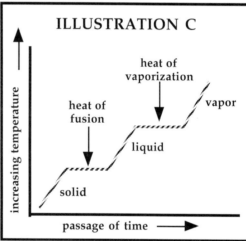

ILLUSTRATION C

increasing temperature →

heat of vaporization

heat of fusion

vapor

liquid

solid

passage of time ⟶

Go!

Have students construct the set-up shown in Figure D on Journal Sheet #3 and complete the activity as directed. Be sure they mark off "time intervals" on the "x" axis of the graph and "temperature readings" (i.e., from 50° C to –10° C) on the "y" axis. Their graph will go down from upper left to lower right, in the opposte direction as shown in Illustration C.

Materials

ring stand and clamps, 500 ml beakers, Celsius thermometers, safety goggles and lab apron, water, ice, test tubes

CH1 Journal Sheet #3

THE PROPERTIES AND PHASES OF MATTER

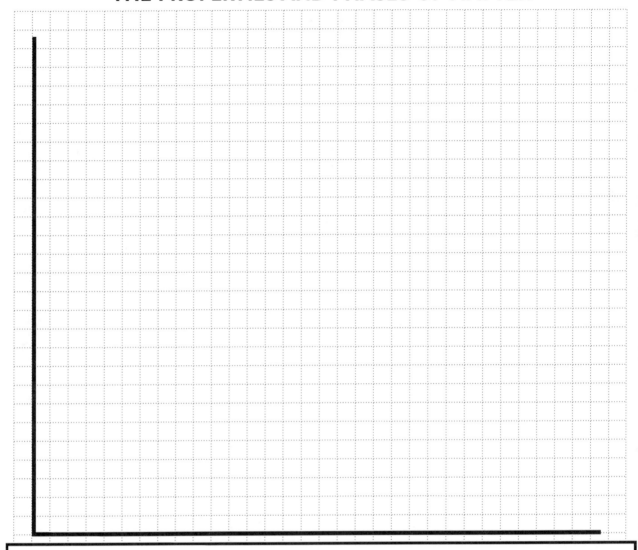

FIGURE D

<u>Directions</u>: (1) Pour 10 ml of water into a test tube. (2) Secure the test tube in a clamp and place it in a 500 ml beaker as shown. (3) Carefully surround the test tube with a mixture of ice and water (mostly ice). (4) Secure a thermometer in a clamp and submerge the thermometer into the test tube as shown. (5) Read and graph the temperature of the water every minute for 20 minutes. (6) Carefully remove the thermometer and test tube. (7) Examine and describe the contents of the test tube.

GENERAL SAFETY PRECAUTIONS

Be careful when cleaning up. Extreme cold can crack and splinter the glass test tube. Wear safety goggles and apron. DO NOT TOUCH BROKEN GLASS. Ask your instructor to assist you in disposing of it.

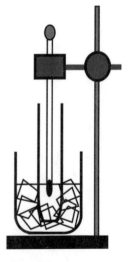

THE PROPERTIES AND PHASES OF MATTER

Work Date: ____/____/____

LESSON OBJECTIVE

Students will explain and demonstrate that evaporation is a "cooling" process.

Classroom Activities

On Your Mark!

Review the material covered in Lesson #3 and add the process called **ionization** to the changes that take place in matter. Explain that superheating a vapor can cause the atoms of the gas to "fall apart" as discussed in the Fact Sheet. An extremely hot gas made of charged or "ionized" particles is formed. Light a match and explain that the atoms caught in the flame are most probably in the form of a **plasma**. The **solar wind** that blows toward earth from the sun is a plasma of ionized particles.

Get Set!

Review the process of vaporization. Ask students: "What causes a liquid to vaporize?" Answer: heat. But what happens to the substance being heated as the hot particles leave? Answer: If the hot particles are leaving, then the substance must be getting cooler. <u>Evaporation is a cooling process</u>. Ask students: "How do you feel when you get out of a swimming pool on a windy day?" Answer: cold. The water evaporating from your skin is taking away the heat it gained from your body.

Go!

Have students construct the set-up shown in Figure E on Journal Sheet #4 and complete the activity as directed. Be sure they mark off "time intervals" on the "x" axis of the graph and "temperature readings" (i.e., from 50° C to −10° C) on the "y" axis. They will discover that alcohol evaporates more quickly than water as the temperature on the "wet bulb" thermometers both drop over time.

Materials

thermometers, cotton or paper towels, water, rubbing (isopropyl) or ethyl alcohol, a text book

CH1 JOURNAL SHEET #4

THE PROPERTIES AND PHASES OF MATTER

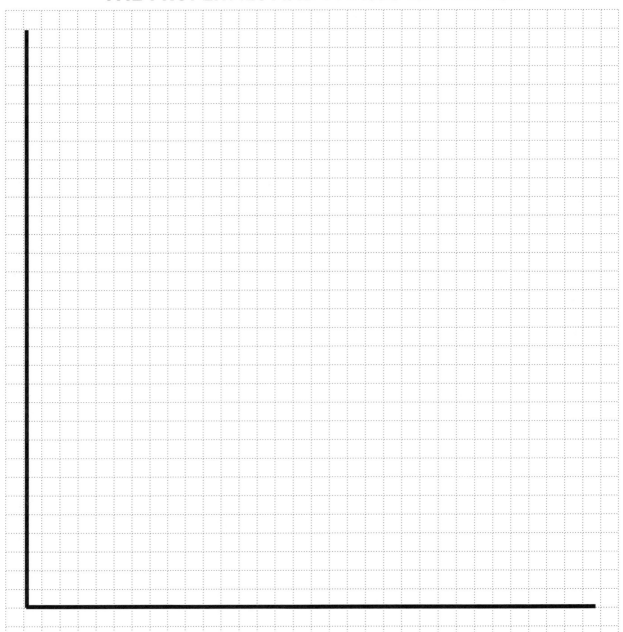

FIGURE E

Directions: (1) Wrap the bulbs of three thermometers in cotton or paper towel. (2) Secure the thermometers in between the pages of a textbook as shown. (3) Place a towel underneath the set up and soak the wrapping around the first thermometer with water. Soak the second wrapping with rubbing alcohol. Leave the wrapping around the third thermometer dry. (4) Read and graph the temperatures of each thermometer every minute for 10 minutes.

CH1 REVIEW QUIZ

Directions: Keep your eyes on your own work.
Read all directions and questions carefully.
THINK BEFORE YOU ANSWER!
Watch your spelling, be neat, and do the best you can.

CLASSWORK	(~40): _____
HOMEWORK	(~20): _____
CURRENT EVENT	(~10): _____
TEST	(~30): _____

TOTAL   (~100): _____
(A ≥ 90, B ≥ 80, C ≥ 70, D ≥ 60, F < 60)

LETTER GRADE: _____

TEACHER'S COMMENTS: _____
_____

THE PROPERTIES AND PHASES OF MATTER

TRUE–FALSE FILL-IN: If the statement is true, write the word TRUE. If the statement is false, change the underlined word to make the statement true. *15 points*

_____ 1. Matter exists in <u>five</u> states or phases.

_____ 2. <u>Liquids</u> have definite shape and definite volume.

_____ 3. <u>Solids</u> can change shape while keeping the same volume.

_____ 4. <u>Plasma</u> changes both shape and volume.

_____ 5. <u>Vapor</u> is a highly energized form of matter made of electrically charged particles like electrons.

_____ 6. The sun's surface is made of <u>plasma</u>.

_____ 7. Matter <u>can</u> change from one phase to another.

_____ 8. Heating a solid causes the atoms that make up the solid to move <u>faster</u>.

_____ 9. As atoms absorb heat energy they move more freely turning a solid into a liquid. This process is called <u>melting</u>.

_____ 10. When the atoms of a liquid absorb more heat they may leave the liquid to form a gas or vapor. This process is called <u>vaporization</u>.

_____ 11. Some solids change directly into a gas when heated. This is called <u>sublimation</u>.

_____ 12. <u>Condensation</u> refers to a gas cooling to form a liquid.

_____ 13. <u>Freezing</u> refers to a liquid cooling to form a solid.

_____ 14. The shape or form of the matter in a <u>chemical</u> change may change while the matter remains the same.

_____ 15. Burning a piece of wood is an example of a <u>physical</u> change.

PROBLEM

Directions: Use Figure A to complete sentences 16 through 20. *15 points*

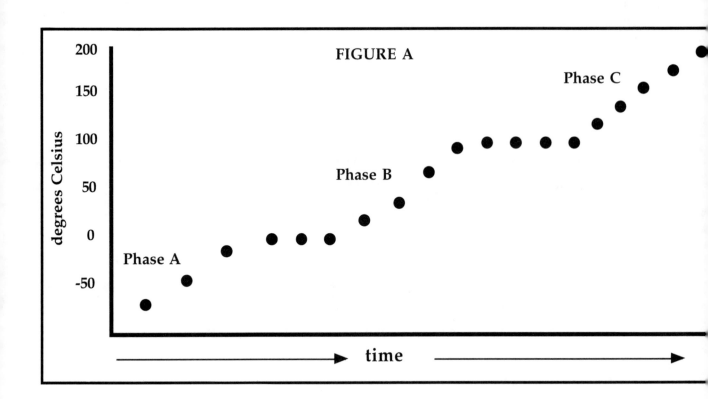

16. During Phase A the substance being heated is probably in the _____ phase.

17. During Phase B the substance being heated is probably in the _____ phase.

18. During Phase C the substance being heated is probably in the _____ phase.

19. Between Phase A and Phase B the substance being heated is probably _____.

20. The substance shown in this graph is probably _____.

_____   _____   ____/____/____
Student's Signature   Parent's Signature   Date

MIXTURES

TEACHER'S CLASSWORK AGENDA AND CONTENT NOTES

Classwork Agenda for the Week

1. Students will discuss the nature of mixtures and begin an experiment that shows that air is a mixture of different gases.

2. Students will identify the various components of a solid mixture.

3. Students will separate the components of a solution by distillation.

4. Students will demonstrate that different substances dissolve at different rates and that temperature affects the solubility of substances.

Content Notes for Lecture and Discussion

The French chemist **Auguste Laurent** (b. 1807; d. 1853) once noted that chemistry is the study of substances that do not exist. What he meant was that the absolute purification of a given substance is not possible to achieve. Even today, when chemists produce a batch of "pure substance" they are mindful of the impurities that contaminate their sample. They express that degree of contamination in "parts per million or billion." Nevertheless, **purity** is a fundamental concept of chemistry, alluding to the homogeneity of a substance and its capacity to yield reproducible effects under given conditions. A contemporary of Auguste Laurent, the French chemist **Michel-Eugéne Chevreul** (b. 1786; d. 1889) determined the first criteria for assessing the purity of a substance. Chevreul established **melting point** and **boiling point** as essential characteristics in the identification of any pure material. Both Laurent and Chevreul worked with a particularily miscible group of chemical substances: the huge family of organic compounds.

Unlike "pure substances," the familiar materials (i.e., wood, plastic, metal, and a vast variety of common solutions) with which we come in contact every day are mixtures that exhibit variable properties. Early metallurgists found it nearly impossible to do systematic chemistry until they perfected methods for extracting "pure" elements and compounds from their ores as well as methods to filter and distill solutions. The discoverer of beet sugar and "father of the sugar industry," German chemist **Andreas Sigismund Marggraf** (b. 1709; d. 1782), set down the first rules for establishing the purity of **chemical reagents**. His methods made it possible for chemistry to become a quantitative study of the properties and interactions of "different forms of matter." After Marggraf, one could truly work with minimally contaminated substances.

In later units, students will learn to distinguish a *mixture* from a *compound*. A **mixture** is a combination of substances that can be separated by ordinary physical means. The properties (i.e., melting point and boiling point) of the individual substances in a mixture do not change upon **physical** separation or recombination. The mixed substances retain their individual identities. A **compound**, on the other hand, undergoes a complete transformation when separated into its individual **elements** by **chemical change**.

In Lesson #1, students will identify some of the mixtures with which they are familiar and begin an experiment that will last the week. They will demonstrate that our atmosphere is a combination of gases by removing one gas, namely oxygen, from a closed system.

In Lesson #2, students will examine and quantify the components of a mixture of solids.

In Lesson #3, students will **distill** a solution of colored water, separating the water from the food coloring. They will learn that distillation is a common method for purifying a variety of useful substances like crude oil.

In Lesson #4, students will test the **solubility** of some simple salts (or sugars). They will discover that the amount of substance that can be dissolved in water is dependent upon the temperature of the water.

ANSWERS TO THE HOMEWORK PROBLEMS

Students will discover that the seemingly solid colored ink in marking pens is sometimes the combination of more than one colored chemical. They will report a separation of the "solid colors" into a rainbow of varied hues. Their particular results, of course, will depend upon the felt tip products they use.

ANSWERS TO THE END-OF-THE-WEEK REVIEW QUIZ

1. rarely (sometimes)
2. physical
3. true
4. glass
5. concrete

6. steel
7. ores
8. true
9. solute
10. solvent

11. true
12. true
13. increases
14. true
15. true

16. X
17. Y
18. 60 grams in 200 ml
19. 70° C
20. Z

CH2 FACT SHEET

MIXTURES

CLASSWORK AGENDA FOR THE WEEK

(1) Demonstrate that air is a combination of gases.
(2) Examine the components of a mixture made of solids.
(3) Separate the components of a solution by distillation.
(4) Show that different substances dissolve at different rates.

Matter rarely exists in a pure state. Most materials are combined with different materials to form *mixtures*. A **mixture** is any combination of substances that can be separated by *physical* means. Mixed matter can be sifted, filtered, or distilled to separate the materials making up the mixture. The substances themselves are not changed when they are separated in this manner. They are simply isolated from one another and purified.

Many of the substances we take for granted every day are manmade mixtures. Glass is a mixture of melted sand (called silica) and soda-salt (called sodium carbonate). Glass was introduced in the Middle East around 3,000 B.C. Concrete is a mixture of sand and broken stones. Concrete was first used as a building material in the Middle East around 700 B.C. Steel has been in existence for about two hundred years. It is an **alloy** which is a mixture of metals. Steel is a combination of iron and carbon. Metals such as iron are found naturally mixed with other metals to form **ores**. Iron ores must be heated to very high temperatures before the pure iron can be extracted from the ore.

A **solution** is a liquid mixture. Solutions contain particles of **solute** dissolved in a liquid **solvent**. Most common liquids are mixtures. Even the water that comes out of the faucet in your home is a mixture of water and a variety of minerals. Scientists classify solutions as either **homogeneous** or **heterogeneous**. A homogeneous solution contains a well dissolved solute present in equal amounts throughout the solvent. Salt water is a homogeneous solution. A heterogeneous solution contain solutes that are not equally distributed throughout the solvent. Salad dressing made with oil and vinegar is a heterogeneous solution.

The amount of a solute that can be dissolved in a liquid solvent depends on the temperature of a solution. Increasing temperature usually increases the amount of material that can be dissolved in the solution. **Solubility** is a measure of the amount of solute that can be dissolved in a solvent at a particular temperature. Scientists use **solubility graphs** to show the solubility of different substances. Solubility can be used to identify an unknown sample of matter.

Gases are also found mixed together. The air we breathe is a combination of gases. Our atmosphere is 78% nitrogen gas and 21% oxygen gas. The remaining 1% is a mixture of other gases like carbon dioxide and argon.

CH2 Fact Sheet (cont'd)

Homework Directions

Many substances can be separated by a technique called **paper chromatography**. Paper chromatography takes advantage of the fact that the molecules of different substances have different weights and will be absorbed onto a paper surface at different speeds. Perform the following experiment to demonstrate how the ink in felt markers can be separated into colors by paper chromatography. Use Illustration A to help you.

1. Gather together several felt colored markers.
2. Cut a coffee filter into strips about three inches long and one-half-inch wide. Set aside one strip for each colored marker.
3. Make a colored spot near the bottom of each strip with a different color.
4. Pour some water into a paper cup or glass.
5. Touch the end of each strip to the surface of the water leaving the colored dot above the surface.
6. Fold the strips over the rim of the cup or glass and secure them with a rubber band.
7. Place the glass in a safe place and record your observations every fifteen minutes for about two hours.

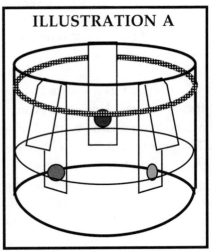

ILLUSTRATION A

Assignment due: _____

_____   _____   ___/___/___
Student's Signature   Parent's Signature   Date

MIXTURES

Work Date: ____/____/____

LESSON OBJECTIVE

Students will discuss the nature of mixtures and begin an experiment that shows that air is a mixture of different gases.

Classroom Activities

On Your Mark!

Show students samples of solid (e.g., granite rocks), liquid (e.g., milk), and gas (e.g., an inflated balloon) mixtures. Begin discussion by asking students the following question: Are these samples pure substances? Answer: No. They are mixtures. Ask them to discuss what makes a sample pure. Lead them to the conclusion that a pure substance is one that "reacts" in exactly the same way under specified conditions. A mixture may not react in the same way every time because its components may not be evenly mixed. Define a **mixture** as any combination of substances that can be separated by ordinary physical means. Explain that in a **physical change** the individual substances in a mixture do not change. They retain their physical properties like melting and boiling points.

Get Set!

Have students make a list of several solid, liquid, and gas mixtures with which they are familiar. Be sure they include the materials that make up each mixture (e.g., What individual substances make up sea water?).

Go!

Assist students in setting up the experiment shown in Figure A on Journal Sheet #1. Explain that oxygen combines readily with a variety of substances. Oxygen can cause an apple to turn brown or iron to rust. The experiment that they will conduct will take advantage of oxygen's ability to combine with other substances. By the end of the week the oxygen inside the test tube will combine with the iron in the steel wool to form rust (iron oxide). Atmospheric pressure will cause the water in the beaker to rise up the cylinder to take the place of the used atmospheric oxygen. If the atmosphere were pure oxygen, then the test tube would fill completely. It does not because the remainder of the atmosphere is made of other gases. Conclusion: Air is a mixture.

Materials

ring stand and clamps, beakers, graduated cylinders, steel wool, detergent, samples of solid and liquid mixtures, balloons (if available)

CH2 JOURNAL SHEET #1

MIXTURES

FIGURE A

<u>Directions</u>: (1) Pour some water into a beaker and place it on a ring stand. (2) Wash and rinse a ball of steel wool with detergent and press it down to the bottom of a graduated cylinder. (3) Invert the cylinder into the beaker and secure it as shown. Make sure the rim of the cylinder is not touching the bottom of the beaker. (4) Keep a record of the water level in the cylinder for the next several days. (5) Write a conclusion that explains your observations.

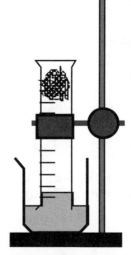

MIXTURES

Work Date: ____/____/____

LESSON OBJECTIVE

Students will identify the various components of a solid mixture.

Classroom Activities

On Your Mark!

Present students with a mixture of iron filings and salt. Ask: How can we separate the iron filings from the salt? They might suggest using a magnet to attract the filings out of the mixture. Or, they might recommend mixing the sample with water then allowing the salt to dissolve before filtering off the iron filings. Both of these methods would work. Tell students to complete the CHALLENGE on Journal Sheet #2.

Get Set!

Present students with any of the following mixtures: bags of "M&M™s," trail mix, mixed nuts, Lego™ blocks, and so on.

Go!

Have students separate the "pure" components from the mixture. They should count the number of pure items in each identified group (i.e., red M&M™s <u>vs</u> brown M&M™s) and graph their results in the space provided on Journal Sheet #2. Have them write the ratios of the components as follows: 20 red M&M™s, 10 brown M&M™s, 20 yellow M&M™s = 2:1:2. Ask students to consider the simple method they used to separate these solids. Did any of the individual components change its identity during the process? Answer: No. This was a physical separation.

Materials

iron filings, salt, bags of candy (see *Get Set!* above)

CH2 JOURNAL SHEET #2

MIXTURES

CHALLENGE

Describe a procedure you could use to separate a mixture of salt and iron filings.

_____

_____

_____

_____

_____

_____

_____

Directions: (1) Physically separate and group the individual components of the mixture given to you by your instructor. (2) Count the number of items in each group of materials. (3) Make a bar graph that shows the relationship between the individual components comprising the mixture.

MIXTURES

Work Date: _____/_____/_____

LESSON OBJECTIVE

Students will separate the components of a solution by distillation.

Classroom Activities

On Your Mark!

Prepare the bent glass tubing shown in Figure B on Journal Sheet #3 before the start of class. Use a Bunsen burner to heat and soften the glass. Then slowly bend the glass into position. Allow the glass to cool and use glycerine to insert the short end of the bent tubing into a single-holed rubber stopper.

Begin class discussion with the definition of a <u>solution</u>. A **solution** is a liquid mixture containing a <u>solute</u> and a <u>solvent</u>. A **solute** is the substance that is dissolved in a liquid. A **solvent** is the liquid in which a substance is dissolved. Ask students to list five familiar solutions (i.e., anything they drink on any given day). Tell them to describe the arrangement of solutes in each solution. For example: Are molecules of sugar solute evenly distributed throughout a can of soda? Does every milliliter of soda contain sugar? If the solute is evenly distributed throughout the solvent, then the solution is a **homogeneous solution**. Is the "orange pulp" solute evenly distributed throughout a solution of orange juice? The pulp may be scattered randomly throughout the solution, but not every milliliter of the juice contains pulp. This type of solution is a **heterogeneous solution**.

Get Set!

Show students the set-up shown in Figure B on Journal Sheet #3. Explain that a more sophisticated set-up like this one is used to separate the solutes in many important solutions. Many of the products we use every day (i.e., plastics, gasoline, heating fuel) all come from a valuable solution called **crude oil**. The solutes in crude oil are separated by the process of **distillation**. Distillation takes advantage of the fact that different solutes have different boiling points. Distillation makes it possible to separate and collect solutes one at a time.

Go!

Allow students ample time to set up and complete the demonstration described in Figure B on Journal Sheet #3.

Materials

ring stand and clamps, Ehrlenmeyer flasks, single-holed rubber stoppers, glass tubing, test tubes, beakers, Bunsen burners, water and food coloring, matches, lab apron

CH2 JOURNAL SHEET #3

MIXTURES

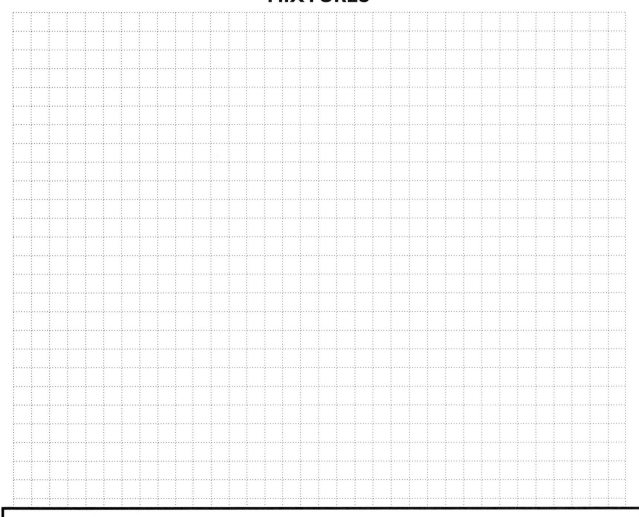

FIGURE B

<u>Directions</u>: (1) Pour 100 ml of water into an Ehrlenmeyer flask. (2) Put a few drops of food coloring into the flask and swirl the flask gently to mix the solution. (3) Place the flask on the ring stand and secure it with a clamp so that it cannot be toppled. (4) Place a test tube in a beaker next to the ring stand. (5) Insert the rubber stopper holding the glass tubing snugly into the flask. DO NOT HOLD OR MANIPULATE THE GLASS TUBING. HANDLE THE RUBBER STOPPER ONLY. Make sure the other end of the glass tubing is inside the test tube. (6) Turn on the Bunsen burner. (7) Record what you observe for the next ten minutes and explain your observations.

GENERAL SAFETY PRECAUTIONS

Be sure you are familiar with the proper use of a Bunsen burner. Wear goggles and lab apron to protect your skin and eyes from being burned by SCALDING HOT STEAM. Do not touch any part of the equipment without heat-resistant gloves or tongs. Clean up when the apparatus is cool.

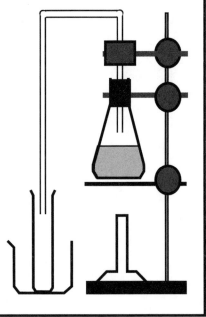

MIXTURES

Work Date: ____/____/____

LESSON OBJECTIVE

Students will demonstrate that different substances dissolve at different rates and that temperature affects the solubility of substances.

Classroom Activities

On Your Mark!

Draw the graph shown in Illustration A on the board. Discuss the meaning of the graph with students. Point out that each line represents the amount of a particular solute that can dissolve in 100 ml of pure water at a given temperature. This amount of water, for example, can only hold about 40 grams of sodium chloride before it becomes **saturated**. No more of that substance will dissolve in that amount of water even at higher temperatures. Ammonium nitrate, on the other hand, is highly **soluble** at low temperatures. Explain that the solubility of a particular solute depends on both the nature of the solute and the temperature of the solvent in which it is being dissolved.

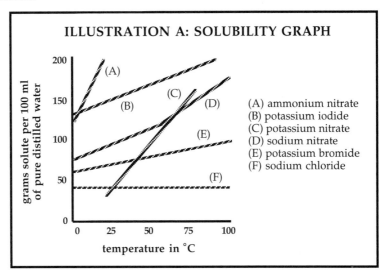

ILLUSTRATION A: SOLUBILITY GRAPH

(A) ammonium nitrate
(B) potassium iodide
(C) potassium nitrate
(D) sodium nitrate
(E) potassium bromide
(F) sodium chloride

grams solute per 100 ml of pure distilled water

temperature in °C

Get Set!

Choose sodium chloride (i.e., common table salt) and, if available, another of the salts indicated in the <u>Solubility Graph</u>. Plain sugar or baking soda will serve just as well for the purposes of this experiment if the other salts are unavailable. Assist students in setting up the apparatus shown in Figure C on Journal Sheet #4.

Go!

Give students time to complete the experiment described in Figure C on Journal Sheet #4. Explain that the results of this experiment **will not match** the graph shown above unless you are using pure distilled laboratory grade water. Plain tap water will not give the same results because it already has minerals dissolved in solution. Nevertheless, students will be able to demonstrate that different substances dissolve at different rates and, time permitting, that temperature affects the solubility of solutes.

Materials

ring stand and clamps, beakers, thermometers, glass stirring rods, hot plates, goggles, heat resistant gloves, water, salt, sugar, baking soda, or any of the salts listed in Illustration A

CH2 JOURNAL SHEET #4

MIXTURES

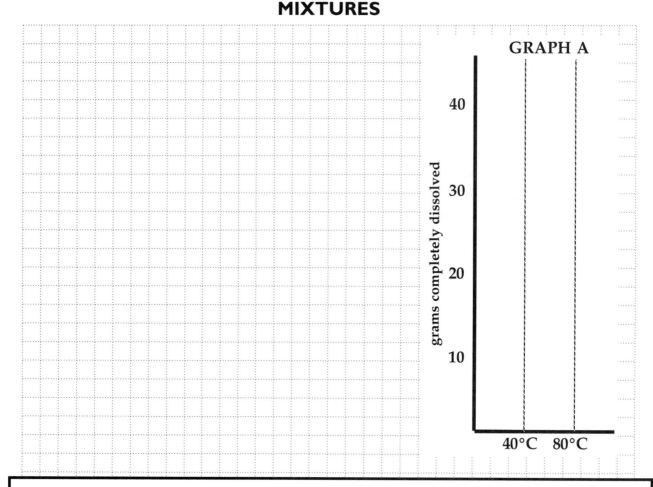

GRAPH A

grams completely dissolved

40

30

20

10

40°C   80°C

FIGURE C

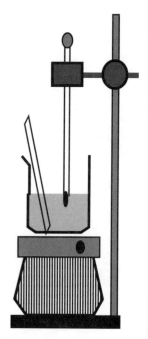

<u>Directions</u>: (1) Pour 100 ml of water into a 500 ml beaker and place it on a hot plate positioned on a ring stand. Secure a thermometer as shown. (2) Turn on the hot plate. (3) Allow the water to heat up to 40° C and adjust the hot plate knob to the lowest setting to maintain the temperature of the water at about 40° C. (4) Add 10 grams of salt (sodium chloride) to the water and stir gently with the stirring rod for one minute. (5) If the salt dissolves completely add another 10 grams. (6) Record on GRAPH A the amount of salt you were able to dissolve completely (10 grams, 20 grams, 30 grams, etc.). (6) Turn off the hot plate and allow the beaker a moment or two to cool. (7) Gently raise the thermometer out of the way and carefully remove the beaker from the hot plate. (8) Rinse the beaker thoroughly and repeat steps #1 through #5 using a second substance given to you by your instructor. (9) If time permits, repeat the experiment again allowing the water to heat up to 80° C.

GENERAL SAFETY PRECAUTIONS

Be sure you are familiar with the proper use of the hot plate. Wear goggles to protect your skin and eyes from being burned by HOT WATER. Do not touch any part of the equipment without heat resistant gloves or tongs. Clean up only when the apparatus is cool.

CH2 REVIEW QUIZ

Directions: Keep your eyes on your own work.
Read all directions and questions carefully.
THINK BEFORE YOU ANSWER!
Watch your spelling, be neat, and do the best you can.

CLASSWORK (~40): _____
HOMEWORK (~20): _____
CURRENT EVENT (~10): _____
TEST (~30): _____

TOTAL (~100): _____
(A ≥ 90, B ≥ 80, C ≥ 70, D ≥ 60, F < 60)

LETTER GRADE: _____

TEACHER'S COMMENTS: _____
_____

MIXTURES

TRUE–FALSE FILL-IN: If the statement is true, write the word TRUE. If the statement is false, change the underlined word to make the statement true. *15 points*

_____ 1. Matter <u>always</u> exists in a pure state.

_____ 2. A mixture is any combination of substances that can be separated by <u>chemical</u> means.

_____ 3. In a <u>physical</u> change, the mixed substances are not changed into other substances when separated.

_____ 4. <u>Concrete</u> is a mixture of melted silica and sodium carbonate.

_____ 5. <u>Glass</u> is a mixture of sand and broken stones.

_____ 6. <u>Iron</u> is an alloy which is a mixture of metals.

_____ 7. Metals like iron are found naturally mixed with other metals to form <u>alloys</u>.

_____ 8. A <u>solution</u> is a liquid mixture.

_____ 9. Salt is the <u>solvent</u> in a solution of salt water.

_____ 10. Water is the <u>solute</u> in a solution of salt water.

_____ 11. A <u>homogeneous</u> solution contains a well dissolved solute present in equal amounts throughout the solvent.

_____ 12. A <u>heterogeneous</u> solution contain solutes that are not equally distributed throughout the solvent.

_____ 13. Increasing temperature usually <u>decreases</u> the amount of material that can be dissolved in the solution.

_____ 14. Our atmosphere is 78% <u>nitrogen</u> gas.

_____ 15. Our atmosphere is 21% <u>oxygen</u> gas.

PROBLEM

Directions: Study the following graph and answer questions 16 through 20. *15 points*

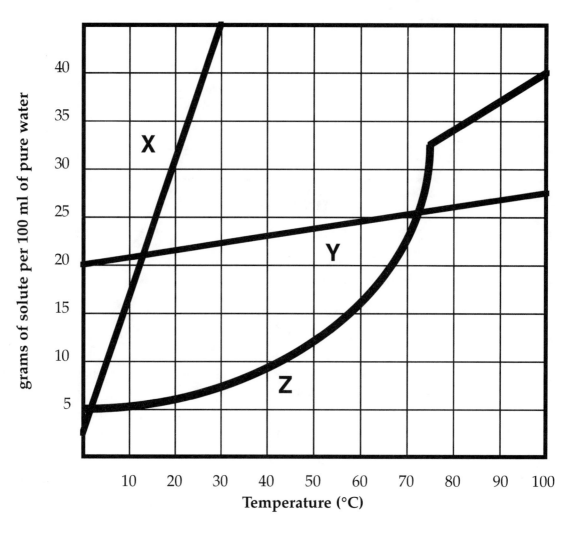

16. Which solute dissolves more quickly than the others? _____

17. Which solute dissolves most slowly at higher temperatures? _____

18. How many grams of solute "X" will completely dissolve in 200 ml of pure water at 20° Celsius? _____

19. At what temperature will 50 grams of solute "Y" be completely dissolved in 200 ml of pure water? _____

20. Which solute does not dissolve at a constant rate in pure water as it is heated? _____

_____   _____   ___/___/___
Student's Signature   Parent's Signature   Date

CRYSTALS

TEACHER'S CLASSWORK AGENDA AND CONTENT NOTES

Classwork Agenda for the Week

1. Students will show how to grow crystals in class and at home.

2. Students will build toothpick and construction paper models of crystals.

3. Students will make accurate drawings of the salt crystals they observe under the microscope or magnifying glass.

4. Students will make accurate drawings of the crystals found in common rocks.

Content Notes for Lecture and Discussion

People have always been attracted to the beauty and variety of crystals. In addition, philosophers such as **Aristotle** (b. 384; d. 322 B.C.) have remarked on the similarities between crystals and plant and animal life. Like plants and animals, crystals grow and exhibit regular geometric forms despite the fact that they are inanimate and solid. The Irish chemist **Robert Boyle** (b. 1627; d. 1691), whose formulation of **Boyle's law** in 1662 explained the relationship between temperature and pressure in a given amount of gas, suggested that molecules in a crystal were clusted into orderly geometric units. The English scientist **Robert Hooke** (b. 1635; d. 1703), best known for popularizing the term "cell" to describe the basic unit of biology, published the first thorough description of different crystalline forms in his book *Micrographia* in 1665. Hooke proposed that the underlying arrangement of the atoms and molecules in a crystal could be determined by examining the orientation of axes and faces of the crystal. For his work, Hooke is considered the "father of crystallography." The French minerologist and founder of modern crystallography, **René-Just Haüy** (b. 1743; d. 1822), developed the first system of classification used to categorize the varied forms of crystals. In 1784, he proposed that the smooth faces of a calcite crystal could be the result of stacking the individually cleaved layers of the crystal. A contemporary of Haüy, the English chemist **William Hyde Wollaston** (b. 1766; d. 1828) invented the first reliable instrument for measuring the angles in crystals. In 1808, Wollaston suggested that an understanding of the three-dimensional arrangement of the atoms in a substance would be of great value to chemists. Wollaston worked diligently in support of the new **atomic theory of matter** proposed by the English chemist **John Dalton** (b. 1766; d. 1844) that very same year in Dalton's book *A New System of Chemical Philosophy*.

In 1895, the German physicist **Wilhelm Röntgen** (b. 1845; d. 1923) discovered X-rays. But it was not until 1912 that father-and-son English physicists, **William** (b. 1862; d. 1942) and **Lawrence** (b. 1890; d. 1971) **Bragg** applied X-rays to the study of crystals. The Braggs devised the first **X-ray spectrometer** used to examine the lattice structure of crystals by X-ray diffraction. Their work—which was the start of **X-ray crystallography**—was based on the idea that X-rays passing through a crystal are diffracted by the orderly arrangement of atoms comprising the crystal.

In Lesson #1, students will begin growing crystals and track the progess of that growth throughout the unit. They will record their observations on a daily basis.

In Lesson #2, students will construct toothpick and construction paper models of crystals in order to visualize the internal arrangement of atoms in each of the six basic types of crystals. This will help students see how the orderly arrangement of atoms gives rise to the crystal's outward appearance.

In Lessons #3 and #4, students will examine salt and rock crystals under a microscope or magnifying glass and make accurate drawings of their observations.

CH3 Content Notes *(cont'd)*

ANSWERS TO THE HOMEWORK PROBLEMS

Students will discover that Epsom salt (hydrated magnesium sulphate) forms crystals with a needle-like appearance. Epsom salt crystals are classified as orthorhombic and have three perpendicular axes of varied lengths.

ANSWERS TO THE END-OF-THE-WEEK REVIEW QUIZ

1. crystals
2. three-dimensional
3. atoms
4. cube
5. Robert Hooke

6. crystallography
7. faces
8. true
9. shape (axes/faces)
10. D-1

11. C-5
12. E-3
13. F-2
14. B-4
15. A-6

CH3 FACT SHEET

CRYSTALS

CLASSWORK AGENDA FOR THE WEEK

(1) Grow crystals.
(2) Construct paper models of different types of crystals.
(3) Examine different salts under the microscope and identify their varied crystalline structures.
(4) Use the microscope to examine crystals found on common rocks.

One of the most interesting and beautiful phenomena in nature is the formation of a crystal. Precious gems such as diamonds, rubies, and emeralds are all crystals. A **crystal** is a three-dimensional structure whose **atoms**, the basic units that make up matter, arrange themselves in an orderly pattern. This pattern repeats itself over and over. Examine Illustration A to see how the atoms of common table salt—made of sodium and chlorine atoms—arrange themselves to form a cube. The shaded spheres in the illustration represent atoms of sodium. The white spheres represent atoms of chlorine. The orderly arrangement of bonded atoms in a crystal of sodium chloride form a cube. The English scientist, **Robert Hooke** (b. 1635; d. 1703) was one of the first to suggest that the geometric patterns seen in a crystal could be used to figure out the arrangement of atoms inside the crystal. Robert Hooke is considered the "father of crystallography." **Crystallography** is the study of crystals. The science of crystallography is crucial to the studies of **minerology** and **chemistry**.

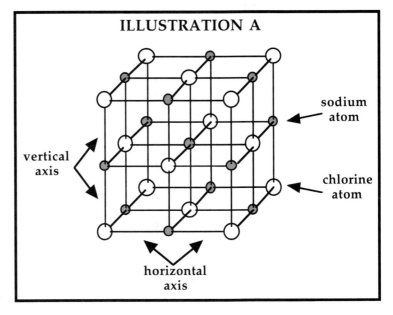

ILLUSTRATION A

sodium atom

chlorine atom

vertical axis

horizontal axis

The orderly arrangement of atoms in a crystal creates smooth surfaces, or **faces**, that help us to imagine the ordering of atoms inside the crystal. Scientists classify crystals according to six basic crystalline arrangements: **regular** (or **cubic**), **tetragonal**, **orthorhombic**, **hexagonal**, **monoclinic**, and **triclinic**. The classification of a crystal as a member of one of these groups depends upon the distances and angles between the atoms in the crystal. Table A on the back of this FACT SHEET shows how crystals are classified according to their shape.

In 1912, scientists discovered that **X-rays** could be used to "see inside" crystals. The technique used to take a picture of the arrangement of atoms in a crystal is called **X-ray crystallography**.

Homework Directions

Perform the following activity to demonstrate how crystals form.

1. Cut a piece of colored construction paper to fit in a small dish.

2. Fill a coffee cup half way with water.

3. Add the sample of Epsom salt given to you by your instructor and stir.

4. Immerse the construction paper into the cup and soak it.

5. Lay the paper out flat in the dish and pour a little more of the solution onto the paper.

6. Set the dish aside in a warm place (not an oven) and allow the water to evaporate for at least one day.

7. Draw a diagram of the crystals that form.

Assignment due: _____

TABLE A: CLASSIFICATION OF CRYSTALS		
Crystal Class	axes	faces
REGULAR CRYSTALS have perpendicular sides of equal length		
TETRAGONAL CRYSTALS have three adjacent perpendicular sides with one long axis		
ORTHORHOMBIC CRYSTALS have three adjacent perpendicular sides with axes of different lengths		
HEXAGONAL CRYSTALS have three horizontal axes of equal lengths (at 120° apart) perpendicular to a longer or shorter fourth axis		
MONOCLINIC CRYSTALS have three axes of varied lengths with two perpendicular and the third slanted		
TRICLINIC CRYSTALS have three axes of different lengths which meet at varied angles		

_____   _____   ___/___/___
Student's Signature   Parent's Signature   Date

CRYSTALS

Work Date: ____/____/____

LESSON OBJECTIVE

Students will show how to grow crystals in class and at home.

Classroom Activities

On Your Mark!

Prepare a sufficient amount of Epsom salt (i.e., 3–4 teaspoons per student) for students who do not have a supply at home with which to complete their home-work assignment. Explain that since their discovery in Epsom, Surrey, England, in 1695, Epsom salts have been used as laxatives and added to warm baths to soothe skin. **Warn students against taking the sample internally.**

Have samples of crystals (common table salt, galena, quartz, gypsum, calcite, etc.) and rocks (granite, marble, gneiss, limestone, quartzite, etc.) available for exam-ination. Ask students to list some of the common characteristics of crystals. Point out that crystals are solid and display a regular geometric appearance with faces and angled corners that repeat from one section of the crystal to another. Allow stu-dents several minutes to handle the samples.

Get Set!

Give a brief lecture about the value of crystals to the study of chemistry using the information in the Teacher's Agenda and Content Notes.

Go!

Have students set-up the experiment shown in Figure A on JOURNAL SHEET #1. Time permitting, begin Lesson #2. Construction of the toothpick and construction paper models requires manual dexterity and may prove time-consuming for some students.

Materials

samples of crystals (common table salt, galena, quartz, gypsum, calcite, etc.) and rocks (granite, marble, gneiss, limestone, quartzite, etc.), Epsom salt, baking soda, 100 ml beakers, paper towels, petri dishes

CH3 JOURNAL SHEET #1

CRYSTALS

FIGURE A

Directions: (1) Pour 50 ml of water into two 100 ml beakers. (2) Pour one teaspoon of baking soda (sodium bicarbonate) into each beaker and stir. Continue to add baking soda as necessary until the solutions are completely saturated. (3) Twist a paper towel tightly into a rope. (4) Immerse one end of the paper towel in each beaker allowing the center of the towel to hang over a petri dish. (5) Draw and record your observations every day at the start of class. (6) At the end of the week, examine the crystals under the microscope and draw your observations. Describe the crystals formed according to one of the six crystal classifications.

Day 1

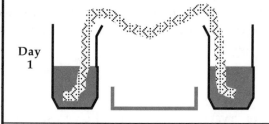

Day 2

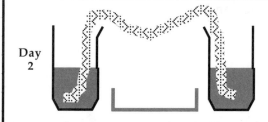

Day 3

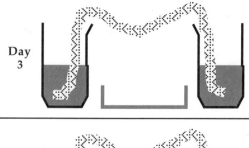

Day 4

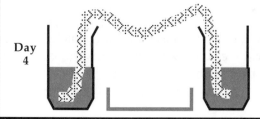

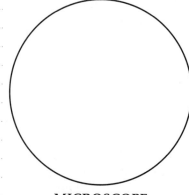

MICROSCOPE
OBSERVATIONS
(Day 4)

CRYSTALS

Work Date: ____/____/____

LESSON OBJECTIVE

Students will build toothpick and construction paper models of crystals.

Classroom Activities

On Your Mark!

Have students examine the experiment set up in Lesson #1. Have them spend a few minutes drawing the extent of crystal growth on Journal Sheet #1.

Review the discussion from Lesson #1. Remind students that as a result of work in X-ray crystallography, we know that the outward appearance of a crystal is the direct result of the orderly arrangement of the atoms that comprise the crystal.

Get Set!

Explain that students can build all of the "toothpick-axis models" and "construction paper-face models" at home for extra credit. But due to class time limitations, each student will build a matching axis and face model of one crystal type. Have students count off from #1 through #6 around the room (i.e., #1, #2, #3, #4, #5, #6, #1, #2, #3, etc.). Students with #1 will build the regular crystal; students with #2 the tetragonal crystals, and so on. Distribute toothpicks (4 or 5 per student) and clay. Show students how to construct a three-dimensional "axis model" of a **regular** or cubic crystal as it appears in Table A on the Fact Sheet. Center two toothpicks at right angles and mold them together with a clay "atom" where they join. Attach four more clay "atoms" to the ends of the toothpicks. Explain that this model represents five atoms in a <u>two-dimensional plane</u> being held together by **electromagnetic forces** (i.e., represented by the toothpicks). Attach a third toothpick and two more clay "atoms" at a right angle to the other toothpicks to represent two more atoms oriented in the <u>third dimension</u>. Explain that the other crystal types shown in Table A on the Fact Sheet will have the toothpicks attached at varying angles. Students will need to cut toothpicks for crystal axes with lengths.

Go!

Instruct students to construct their axis models as shown in Table A in the Fact Sheet. Then, they should build their construction paper models using the templates on Journal Sheet #2. When they are done, they should compare the toothpick models to the construction paper models. Time permitting, group students with the same crystal types and have them "clay" the toothpick models together by adding "clay atoms" and "glue" their paper models together by matching faces.

Materials

toothpicks, clay, tracing or thin white paper, scissors, contruction paper, glue

CH3 JOURNAL SHEET #2

CRYSTALS

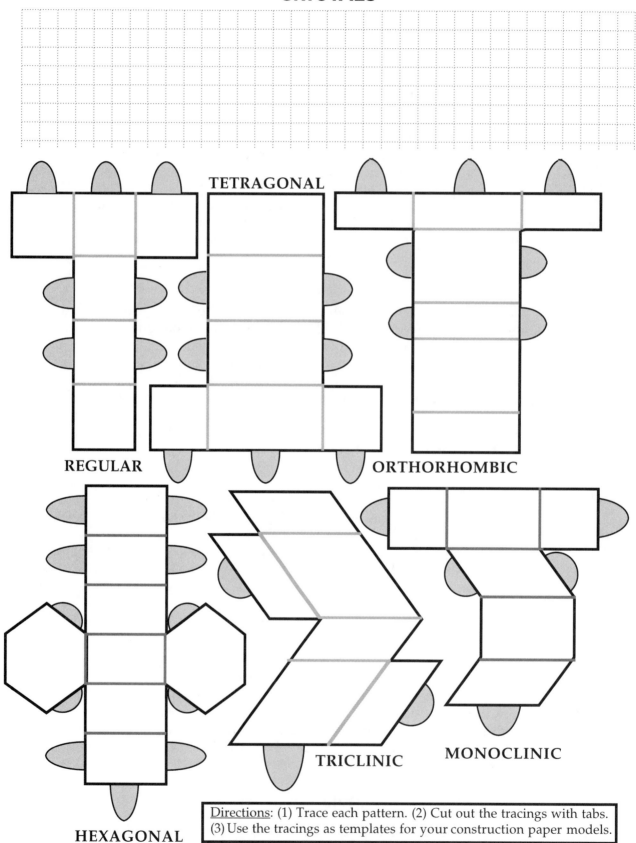

TETRAGONAL

REGULAR

ORTHORHOMBIC

HEXAGONAL

TRICLINIC

MONOCLINIC

Directions: (1) Trace each pattern. (2) Cut out the tracings with tabs.
(3) Use the tracings as templates for your construction paper models.

CRYSTALS

Work Date: ____/____/____

LESSON OBJECTIVE

Students will make accurate drawings of the salt crystals they observe under the microscope or magnifying glass.

Classroom Activities

On Your Mark!

Have students examine the experiment set up in Lesson #1. Have them spend a few minutes drawing the extent of crystal growth on Journal Sheet #1.

Review the proper use of a microscope if you are not using magnifying lenses.

Get Set!

Distribute samples of common table salt (sodium chloride), baking soda (sodium bicarbonate), and other salts you may have in your chemistry cabinet (i.e., potassium chloride, calcium carbonate, potassium nitrate, etc). **Read the warning labels on each sample bottle** to make sure students are aware of the toxicity of your sample substances. Instruct them to wash their hands when they are finished making their observations.

Go!

Have students make accurate drawings of the salt crystals as seen under the microscope or magnifying lens in the spaces provided on Journal Sheet #3. Have them compare their drawings to the crystal types shown in Table A on the Fact Sheet and name the class to which each salt may belong.

Materials

salts (see *Get Set!* above), magnifying lenses or microscopes

CH3 JOURNAL SHEET #3

CRYSTALS

Name of substance:

_____

Class of crystals:

_____

Name of substance:

_____

Class of crystal:

_____

Name of substance:

_____

Class of crystal:

_____

<u>Directions</u>: (1) Use a microscope or magnifying glass to help you examine the crushed salts or other substances given to you by your instructor. (2) Make accurate drawings of your observations and try to determine the correct crystal classification for each sample.

CRYSTALS

Work Date: ____/____/____

LESSON OBJECTIVE

Students will make accurate drawings of the crystals found in common rocks.

Classroom Activities

On Your Mark!

Have students examine the experiment set up in Lesson #1. Have them spend a few minutes drawing the extent of crystal growth and use a microscope or magnifying glass to draw their <u>microscope observations</u> on Journal Sheet #1.

Review the proper use of a microscope if you are not using magnifying lenses.

Get Set!

Distribute samples of rock (i.e., granite, marble, gneiss, limestone, quartzite, etc.). Show students how to use a dissecting needle to scrape small crystals off the rock for viewing under the microscope or magnifying glass.

Go!

Have students make accurate drawings of the rock crystals as seen under the microscope or magnifying lens in the spaces provided on Journal Sheet #3. Have them compare their drawings to the crystal types shown in Table A on the Fact Sheet and name the class to which each salt may belong.

Materials

rocks (see *Get Set!* above), magnifying lenses or microscopes, dissecting needles

CH3 JOURNAL SHEET #4

CRYSTALS

Type of rock:

_____

Number of different
crystals:

_____

Type of rock:

_____

Number of different
crystals:

_____

Type of rock:

_____

Number of different
crystals:

_____

Directions: (1) Use a microscope or magnifying glass to help you examine the crushed sample of rock given to you by your instructor. (2) Make accurate drawings of your observations and try to determine the number of different classes of crystal in each sample.

CH3 REVIEW QUIZ

Directions: Keep your eyes on your own work.
Read all directions and questions carefully.
THINK BEFORE YOU ANSWER!
Watch your spelling, be neat, and do the best you can.

CLASSWORK   (~40): _____
HOMEWORK   (~20): _____
CURRENT EVENT   (~10): _____
TEST   (~30): _____

TOTAL   (~100): _____
(A ≥ 90, B ≥ 80, C ≥ 70, D ≥ 60, F < 60)

LETTER GRADE: _____

TEACHER'S COMMENTS: _____

_____

CRYSTALS

TRUE–FALSE FILL-IN: If the statement is true, write the word TRUE in the space on the left. If the statement is false, write a word or phrase in the space on the left that replaces the underlined word to make the statement true. *18 points*

_____ 1. Precious gems like diamonds, rubies, and emeralds are all <u>easily found</u>.

_____ 2. A crystal is a <u>two-dimensional</u> structure.

_____ 3. <u>Molecules</u> are the basic units that make up matter.

_____ 4. The orderly arrangement of bonded atoms in a crystal of sodium chloride forms a <u>pyramid</u>.

_____ 5. <u>Sir Isaac Newton</u> is considered the "father of crystallography."

_____ 6. <u>Minerology</u> is the study of crystals.

_____ 7. The orderly arrangement of atoms in a crystal creates smooth <u>curves</u> that help us to imagine the ordering of atoms inside the crystal.

_____ 8. In 1912, scientists discovered that <u>X-rays</u> could be used to "see inside" crystals.

_____ 9. Scientists classify crystals according to their <u>color</u>.

PROBLEM

Directions: Place the letter of the axis arrangement and the number of the face shape that matches the correct crystal classification. *12 points*

CRYSTAL	AXIS LETTER	FACE NUMBER
10. REGULAR	_____	_____
11. TETRAGONAL	_____	_____
12. ORTHORHOMBIC	_____	_____
13. HEXAGONAL	_____	_____
14. MONOCLINIC	_____	_____
15. TRICLINIC	_____	_____

AXES

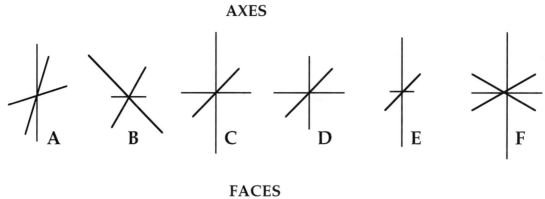

FACES

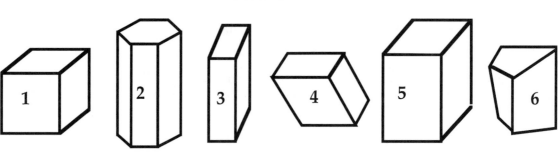

Student's Signature

Parent's Signature

___/___/___
Date

HEAT AND ENERGY TRANSFER

Teacher's Classwork Agenda and Content Notes

Classwork Agenda for the Week

1. Students will be able to explain the difference between heat and temperature and show how heat energy is conducted in a solid.
2. Students will show how heat energy is convected through a liquid or gas.
3. Students will show how heat radiation can be concentrated at a focal point.
4. Students will measure the number of calories in a soda cracker.

Content Notes for Lecture and Discussion

Early philosophers led by **Aristotle** (b. 384 B.C.; d. 322 B.C.) considered "heat" or "hotness" to be an irreducible quality of matter. Heat was one of the four basic qualities—along with earth, air, and water—which in varied combination comprised all material substances. Like most of Aristotle's views, this conception of heat as a primary element of matter persisted until challenged during "The Renaissance of the 17th century." The French mathematician-philosopher **René Descartes** (b. 1596; d. 1650) was one of the first to distinguish between "particles of fire" and heat generated from the "motion" of particles. But it was largely due to the work of **Sir Isaac Newton** (b. 1642; d. 1727) that heat was given formal consideration as a "manner of motion (i.e., mechanical energy)" and not a substance. Nevertheless, the idea that heat was a material substance gave rise to the **caloric theory** which held sway until the 19th century.

Improvements in thermometer technology made it possible for the Scottish chemist **Joseph Black** (b. 1728; d. 1799), who discovered carbon dioxide, to accurately measure the **specific heat capacity** of different substances. That is, the amount of heat energy that a substance could "store." Using a newly invented mercury thermometer, Black defined **specific heat** as the amount of energy required to raise the temperature of a substance by 1°C. A specific increase in the temperature of a piece of iron, for example, required less energy than the same temperature increase in an equal amount of water. Iron is a much better conductor of heat. Black also discovered that the temperature of ice remained constant as it melted despite the fact that it was absorbing energy throughout the process. He reasoned that the added energy must be combining with the particles of water thereby becoming hidden. He called this "concealed energy" **latent heat**. It was not until the work of Black's student, **James Watt** (b. 1736; d. 1819) and the discoveries of the American-English inventor **Benjamin Thompson Rumford** (b. 1753; d. 1814) that the notion of heat as a form of kinetic energy became a widely accepted alternative to the caloric theory.

The study of the kinetic theory of gases by the Austrian physicist **Ludwig Eduard Boltzmann** (b. 1844; d. 1906) bolstered the notion of heat as a manner of motion and furthered the study of the **Second Law of Thermodynamics**: the **Law of Entropy**. Boltzmann was the first to develop a statistical method for describing the behavior of the atoms or molecules in a gas at a specific temperature. His work was used by **Max Planck** (b. 1858; d. 1947) to invent quantum theory and advance the study of electromagnetism. Planck's work allowed scientists to detemine the specific wavelengths and frequencies of radiated energy across the electromagnetic spectrum.

Ironically, basic research on heat in the latter half of the 20th century has been in the area of **cryogenics**: the study of extreme cold.

CH4 Content Notes (cont'd)

In Lesson #1 students will discover the difference between heat and temperature and demonstrate how heat from a chemical flame is transferred through a metal bar by **conduction**. They will also practice temperature conversion problems and change temperature readings from one temperature scale to another.

In Lesson #2, students will show how heat energy is convected through a liquid or gas and discuss the meaning of specific heat.

In Lesson #3, students will show how heat radiation can be concentrated at a focal point. They will draw a simple **Bohr Model** of an atom and discuss how electromagnetic radiation originates inside atoms and is transferred from one region of space to another even through a vacuum.

In Lesson #4, students will construct a "soda can calorimeter" and measure the number of calories in a soda cracker.

ANSWERS TO THE HOMEWORK PROBLEMS

	°F	°C	K		°F	°C	K
1.	59	15	288	6.	50	10	283
2.	41	5	278	7.	167	75	440
3.	68	20	293	8.	−184	−120	153
4.	23	−5	268	9.	−459	−273	0
5.	150	60	333	10.	32	0	273

ANSWERS TO THE END-OF-THE-WEEK REVIEW QUIZ

1. caloric
2. true
3. cannons
4. more
5. true

6. atomic-molecular
7. true
8. true
9. temperature
10. average kinetic energy

11. true
12. true (or Food Calories)
13. 0°C
14. 1,000
15. true

16. true
17. convection
18. radiation
19. true
20. water

21. The air inside the bulb would expand and force the *liquid in the tube downward*.

22. The air in the lower flask would expand and escape through the opening; but the air pressure due to increased temperature in the flask would leave the *level in the tube unaffected*.

23. The air inside the bulb would contract and atmospheric pressure in the flask would force the *liquid in the tube upward*.

24. The air in the lower flask would contract and atmospheric pressure would force air into the opening, leaving the *level in the tube unaffected*.

25. During the cold of winter, the air in the bulb would tend to contract; so, and atmospheric pressure in the flask would *force the liquid in the tube to a higher level* than one might expect in summer.

CH4 FACT SHEET

HEAT AND ENERGY TRANSFER

CLASSWORK AGENDA FOR THE WEEK

(1) Explain the difference between heat and temperature and show how heat energy is conducted in a solid.
(2) Show how heat energy is convected through a liquid and gas.
(3) Show how heat radiation can be concentrated at a focal point.
(4) Measure the number of calories in a soda cracker.

The ancient Greeks believed that heat was a fluid which they called **caloric**. They imagined caloric to be a weightless substance, travelling from hot objects to cold ones. From ancient times to the start of the 19th century, the **caloric theory** proved valuable in explaining many different phenomena. Some philosophers and scientists, however, were not at all convinced of this idea. They argued that heat could also arise from the motion of the particles that made up matter. After all, when your hands are cold you can rub them briskly together to make them warm. Where does this heat come from? How does friction produce heat? Following the invention of the steam engine—and the improvement in its design by the Scottish engineer **James Watt** (b. 1736; d. 1819) in 1782—scientists became more interested in the mechanical aspects of heat and heat transfer.

In 1798, the American-English inventor, **Benjamin Thompson Rumford** [also known as Count Rumford] (b. 1753; d. 1814) was working to develop a more efficient way of manufacturing cannons for the Bavarian military. He was amazed at the incredible amount of heat produced during the boring of a cannon. Rumford published a book explaining that heat was produced by the motion of vibrating particles. Rubbing surfaces together, Rumford explained, caused the particles to collide and vibrate more vigorously. Thus heat was a form of "mechanical energy" that could be transferred from one place to another by colliding particles.

Today, we know that all matter is made up of particles called **atoms** that are always in constant motion. This idea is called the **atomic-molecular theory of matter**. The **internal energy** of an object is the total energy of all of the particles comprising the object as they vibrate and move about. Since it is impossible to directly measure the exact speed and direction of every particle in a substance, scientists refer to the **average kinetic energy** of the particles. **Kinetic energy** is energy of motion. **Temperature** is a measure of the average kinetic energy in the atoms and molecules of a substance.

A **thermometer** is a tool used to measure the temperature of a substance which is the average kinetic energy of the particles that comprise it. Illustration A shows that a thermometer—like all other matter—is made of atoms. The glass walls of the thermometer are made of atoms as is the fluid inside the thermometer tube. The substance outside the thermometer is also made of atoms that are in constant motion. If the atoms surrounding the thermometer absorb energy they will move faster and have more momentum. They will transfer their energy to other atoms with every collision. The glass atoms will absorb that energy and collide with the atoms in the thermometer fluid. As the atoms of fluid move faster, the fluid will expand. Scientists mark the glass tube and read the fluid expansion in units called **degrees**. In the unit on *The Properties and Phases of Matter*, you learned that there are three temperature scales in use today: the **Fahrenheit** scale, the **Celsius** scale, and the **Kelvin** scale. All temperature scales measure the average kinetic energy of the particles in matter. Heat energy can also be expressed in **calories**. One calorie is the amount of energy needed to raise the temperature of pure water 1° Celsius. One thousand calories is equal to 1 Food calorie.

Heat energy can be transferred in three different ways. Through a solid, the heat transfer by colliding particles is called **conduction**. In a fluid, either liquid or gas, the heat transfer by colliding particles is called **convection**. Energy is also given off by the rapid motion of electrons moving around inside atoms. This type of energy is called **radiation**. It is a form of low frequency electromagnetic energy.

Homework Directions

Change the following temperature readings to temperature readings on the other temperature scales as follows: Example: 25°C = 77°F = 350K. Use the formulas on Journal Sheet #1 to help you.

1. 59°F	2. 41°F	3. 68°F	4. 23°F	5. 60°C
6. 10°C	7. 75°C	8. −120°C	5. 0K	10. 273K

Assignment due: _____

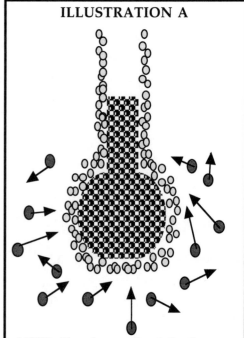

ILLUSTRATION A

NOTE: The glass walls of the thermometer, the fluid inside the thermometer, and the substance outside the thermometer are all made of atoms that are in constant motion.

Student's Signature	Parent's Signature	___/___/___ Date

HEAT AND ENERGY TRANSFER

Work Date: ____/____/____

LESSON OBJECTIVE

Students will be able to explain the difference between heat and temperature and show how heat energy is conducted in a solid.

Classroom Activities

On Your Mark!

Begin with a review discussion of the arrangement of atoms in a crystal. The atoms are held together in an orderly arrangement that gives the crystal a definite shape: Ask students to consider what would happen to the atoms if they began to absorb energy (i.e., from an open flame). Point out that atoms would move about more vigorously, slamming into one another with greater momentum. This could change the arrangement of atoms in the crystal and alter its shape. The crystal could melt and eventually vaporize. Explain that this change was the result of an increase in the momentum of the atoms in the system. The transfer of energy from atom to atom in this solid is called **conduction**. Ask students to consider what would happen if a thermometer—like the one shown on the back of the Fact Sheet—was placed against the crystal. Have students analyze the system at the atomic level and write a brief statement that explains why the liquid inside the thermometer reads higher and higher degrees as the crystal is warmed. The explanation is in paragraph #4 of the Fact Sheet. Guide students to the conclusion that a thermometer does not measure heat energy directly. Instead, it reflects the **average kinetic energy** of the atoms in a system. **Heat energy** is a measure of the total energy (i.e., potential energy plus kinetic energy) of a system. The heat energy released during a chemical change in a substance can be measured using a **calorimeter**. The unit of heat energy is the **calorie**. One calorie is the amount of energy needed to raise the temperature of 1 gram of pure water one degree Celsius.

Get Set!

Distribute thermometers that read both degrees Fahrenheit and degrees Celsius. Remind students that each thermometer employs a different scale to read the same amount of average kinetic energy (i.e., $32°F = 0°C = 273K$). Show students how to use the temperature conversion formulas on Journal Sheet #1 to change temperature readings from Fahrenheit to Celsius to Kelvin, etc.

EXAMPLES:

To change 35°C to °F:

$$°F = \frac{9}{5}(°C) + 32$$

$$°F = \frac{9}{5}(35°) + 32$$

$$°F = 63° + 32$$

$$°F = 95°$$

To change 77°F to °C:

$$°C \quad \frac{5}{9}(°F - 32)$$

$$°C \quad \frac{5}{9}(77° - 32)$$

$$°C \quad \frac{5}{9}(45°)$$

$$°C \quad 25°$$

To change °C to Kelvins: add 273. To change °F to Kelvins: find °C then add 273.

Go!

Have students construct the set-up shown in Figure A on Journal Sheet #1 and perform the activity.

Materials

ring stands and clamps, heat-resistant gloves, metal bars from ring stands, thermometers, Bunsen burners

CH4 JOURNAL SHEET #1

HEAT AND ENERGY TRANSFER

Temperature Conversions

To change °C to °F:

$$°F = \frac{9}{5}(°C) + 32$$

(1) Fill in degrees Celsius and multiply by nine-fifths.
(2) Add 32.

To change °F to °C:

$$°C = \frac{5}{9}(°F - 32)$$

(1) Fill in degrees Fahrenheit and subtract 32.
(2) Multiply by five-ninths.

To change °C to K:

$$K = °C + 273$$

FIGURE A

Directions: (1) Construct the set-up shown making sure that the thermometer is in contact with metal rod. (2) Light the Bunsen burner. (3) Record the temperature readings every 15 seconds for 2 minutes.

time	temperature reading
_____	_____
_____	_____
_____	_____
_____	_____
_____	_____

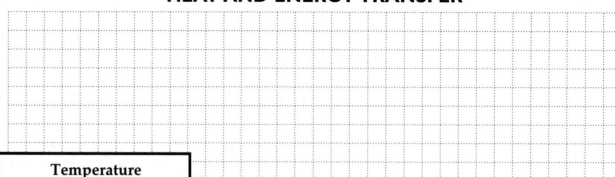

GENERAL SAFETY PRECAUTIONS

Be sure you are familiar with the proper use of a Bunsen burner. Wear goggles to protect your skin and eyes. Do not touch any part of the equipment without heat resistant gloves or tongs. Clean up when the apparatus is cool.

HEAT AND ENERGY TRANSFER

Work Date: ____/____/____

LESSON OBJECTIVE

Students will show how heat energy is convected through a liquid or gas.

Classroom Activities

On Your Mark!

Before the start of class prepare the 1-hole and 2-hole rubber stopper and glass tubing assemblies shown in Figure B and Figure C on Journal Sheet #2.

Review the results of the activity completed in Lesson #1 and repeat the distinction between **heat energy** and **temperature**. Explain that the metal bar used in Lesson #1 did not hold heat very well but transferred the heat very quickly. In other words, the metal was a great conductor of heat but did not have the capacity to "hold" or store the heat. Scientists can measure the capacity of a substance to hold heat. The capacity of a substance to store chemical energy is called **specific heat**. For example, water has a specific heat equal to "1" because—by definition—it takes one calorie of energy to raise the temperature of water 1°C. The specific heat of iron, on the other hand, is only 0.11. That is, it only takes 0.11 calories to raise the temperature of iron 1°C. That's only about one-tenth the amount of energy needed to raise the temperature of an equal amount of water. Table A lists the specific heat of some familiar substances. Ask students to discuss the factors that would make a substance a good **insulator**. Would such a substance have a low or high specific heat? Answer: Insulators have a high specific heat. They store heat energy and prevent its transfer.

TABLE A	
Specific heat* of familiar substances	
air	0.24
aluminum	0.22
brass	0.09
copper	0.09
glass	0.20
gold	0.03
ice	0.50
iron	0.11
lead	0.03
silver	0.06
water	1.00
zinc	0.09
*approximated to nearest hundredths place	

Get Set!

Redefine **conduction** as the transfer of energy through a solid. Point out that energy can also be transferred from one particle to another in fluids. This type of energy transfer is called "convection." Define **convection** as the transfer of energy through a fluid (i.e., liquid or vapor).

Go!

Have students complete the two activities described on Journal Sheet #2. In the first activity (shown in Figure B) they will observe the warmed water escaping from the flask when it is placed in the large beaker or bucket of cold water. In the second activity (shown in Figure C) they will observe the colored water rise in the tube as the air in the flask expands due to heat convection.

Materials

heat-resistant gloves, goggles, Ehrlenmeyer flasks, flat-bottomed Florence flasks, large beakers or buckets, hot plates, two-holed rubber stoppers, one-holed rubber stoppers, glass tubing, food coloring, water

CH4 JOURNAL SHEET #2

HEAT AND ENERGY TRANSFER

FIGURE B

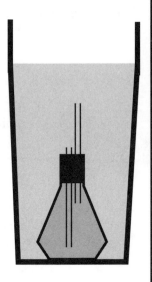

Directions: (1) Fill a small Ehrlenmeyer flask with water and several drops of food coloring. (2) Stopper the flask with the two-holed rubber stopper provided by your instructor. (3) Warm the flask on a hot plate on a low setting for 2 minutes. (4) While the flask is warming, fill a large beaker or bucket with cold water. (5) Use tongs or heat-resistant gloves to transfer the flask to the bottom of the larger beaker or bucket. USE EXTREME CAUTION in completing this step. Although laboratory glassware is tempered to withstand drastic temperature changes there is always the possibility that the glassware will shatter. Follow the GENERAL SAFETY PRECAUTIONS. (6) Record your observations.

GENERAL SAFETY PRECAUTIONS

Be sure you are familiar with the proper use of the hot plate. Wear goggles to protect your skin and eyes from being burned by HOT WATER. Do not touch any part of the equipment without heat-resistant gloves or tongs. Clean up when the apparatus is cool.

FIGURE C

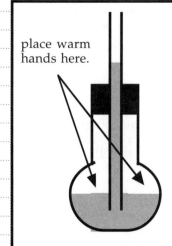

place warm hands here.

Directions: (1) Pour colored water into a flask as shown and stopper the flask with the one-holed rubber stopper given to you by your instructor. (2) Rub your hands together and place them over the flask as shown. (3) Record your observations. (4) Rub ice over the flask in the same place you positioned your hands. (5) Record your observations.

HEAT AND ENERGY TRANSFER

Work Date: ____/____/____

LESSON OBJECTIVE

Students will show how heat radiation can be concentrated at a focal point.

Classroom Activities

On Your Mark!

Review the distinction between **conduction** and **convection**. Ask students to consider the following: Can energy be transferred from one place to another through a space that has no matter (i.e., through a vacuum)? Answer: Yes. Energy can be transferred by means of **radiation**. One form of radiation is **electromagnetic radiation**. Explain that electromagnetic radiation is produced by **electrons** changing their position inside atoms. Draw Illustration A on the board and have students copy the drawing on to Journal Sheet #3. How scientists came to decide on this particularly useful model of the atom will be discussed in another unit.

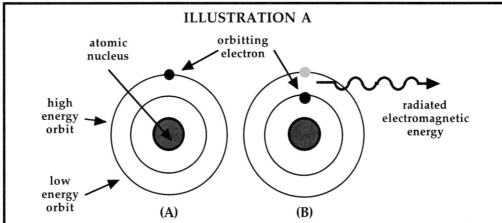

ILLUSTRATION A

As the electron jumps from a high energy orbit in "A" to a low energy orbit in "B" it loses energy to the environment. This energy escapes from the atom as radiated electromagnetic energy. The smaller the jump, the less energy is radiated. This radiated energy can be absorbed by the atoms in distant matter causing them to vibrate more vigorously (i.e., increase their kinetic energy).

Get Set!

Ask students to consider how energy gets from the sun to earth. Have students copy the following list of the different frequencies of electromagnetic radiation. The list is in order of increasing energy: **radio waves**, **microwaves**, **infrared**, **visible light** (colors), **ultraviolet**, **x-rays**, **gamma rays**. Explain that infrared radiation is the energy that "warms" the air of our atmosphere, the water in the oceans, and solid matter all around us. Like visible light, infrared energy can be "bent" and focussed using a magnifying glass.

Go!

Have students perform the activity described in Figure D on Journal Sheet #3.

Materials

thermometers, sun or flashlight, magnifying glass

CH4 JOURNAL SHEET #3

HEAT AND ENERGY TRANSFER

FIGURE D

Directions: (1) Place two thermometers on a paper towel. (2) If it is a sunny day, focus the sun's rays on the bulb of one of the thermometers with a hand held magnifying glass. Or, use a flashlight. (3) Record the thermometer readings from both thermometers every 30 seconds for 5 minutes.

HEAT AND ENERGY TRANSFER

Work Date: ____/____/____

LESSON OBJECTIVE

Students will measure the number of calories in a soda cracker.

Classroom Activities

On Your Mark!

Review the definition of a **calorie** as the amount of energy needed to raise the temperature of one gram of pure water one degree Celsius. Raising the temperature of 100 grams (i.e., 100 ml) of water one degree Celsius would therefore require 100 calories of energy.

Get Set!

Point out that one **Food Calorie** is equal to 1,000 calories as measured using a thermometer or calorimeter. Have students read the ingredients label on a soda cracker box paying particular attention to the **calories per serving information**. Example: "Calories per serving = 140; serving size = 10 crackers." Have students calculate the number of Food Calories in one cracker. In the example above this is equal to 14 Food Calories (or 14,000 calories) per cracker. Burning one cracker—and using <u>all</u> of the released energy to heat the water in a soda can— would warm 100 ml of water about 140 degrees Celsius. This is quite a lot of heat! However, in the set-up shown in Figure E on Journal Sheet #4 most of the heat from the burning cracker will escape.

Go!

Assist students in performing the activity described in Figure E on Journal Sheet #4. In a real calorimeter all of the energy is captured so the rise in temperature caused by the total burning of a substance can be measured with accuracy. After determining the rise in temperature caused by the burning cracker (i.e., the number of calories from the cracker that went into warming the water) have students calculate the efficiency of their "soda can calorimeter." They can do this by dividing their measured results by the real calorie measure derived from the ingredients label. Example: a 35 degree rise in 100 ml water required 3,500 calories. If the actual calories per cracker derived from the ingredients label was 14,000 calories per cracker, then the device is only 25% efficient. 75% of the heat from the cracker escaped.

Materials

ring stands and clamps, ring clamps, thermometers, heat-resistant gloves, dissecting needles, soda cans, water, soda crackers

CH4 JOURNAL SHEET #4

HEAT AND ENERGY TRANSFER

FIGURE E

<u>Directions</u>: (1) Pour 100 ml of water from a beaker into a soda can. (2) Secure the soda can with two ring clamps as shown. (3) Lower a thermometer into the can just below the surface of the water. (4) Skewer a soda cracker at the end of a dissecting needle and secure the needle with a clamp as shown. (5) Record the temperature of the thermometer. (6) Light a match to the soda cracker until the cracker burns on its own. (7) Record the temperature reading when the cracker is completely burned. (8) Since one calorie is the amount of energy needed to raise the temperature of water one degree Celsius, the number of calories in the cracker is equal to the rise in tem-perature caused by the burning cracker x100 (i.e., the volume of water in the can).

rings

dissecting
needle

GENERAL SAFETY PRECAUTIONS

Wear goggles to protect your skin and eyes when working with a flame. Do not touch any part of the equipment without heat-resistant gloves or tongs. Clean up only when the apparatus is cool.

CH4 REVIEW QUIZ

Directions: Keep your eyes on your own work.
Read all directions and questions carefully.
THINK BEFORE YOU ANSWER!
Watch your spelling, be neat, and do the best you can.

CLASSWORK   (~40): _____
HOMEWORK   (~20): _____
CURRENT EVENT   (~10): _____
TEST   (~30): _____

TOTAL   (~100): _____
(A ≥ 90, B ≥ 80, C ≥ 70, D ≥ 60, F < 60)

LETTER GRADE: _____

TEACHER'S COMMENTS: _____

_____

HEAT AND ENERGY TRANSFER

TRUE–FALSE FILL-IN: If the statement is true, write the word TRUE in the space on the left. If the statement is false, write a word or phrase in the space on the left that replaces the underlined word to make the statement true. *20 points*

_____ 1. The ancient Greeks believed that heat was a fluid which they called <u>parboil</u>.

_____ 2. With the improvement of the steam engine by Scottish engineer <u>James Watt</u>, scientists became more interested in the mechanical aspects of heat and heat transfer.

_____ 3. Count Rumford developed his theory of heat while manufacturing <u>computers</u>.

_____ 4. Rumford explained that particles collide and vibrate <u>less</u> vigorously as the result of friction.

_____ 5. Today, we know that all matter is made up of particles called <u>atoms</u> that are always in motion.

_____ 6. The idea that all matter is made of tiny particles that are in constant motion is called the <u>Rumford</u> theory.

_____ 7. The internal energy of an object is the total energy of <u>all</u> of the particles comprising the object.

_____ 8. Kinetic energy is energy of <u>motion</u>.

_____ 9. <u>Heat</u> is a measure of the average kinetic energy in the atoms and molecules of a substance.

_____ 10. A thermometer is a tool used to measure the <u>heat content</u> of the particles in a substance.

_____ 11. Scientists measure temperature in units called <u>degrees</u>.

_____ 12. Scientists measure food energy in units called <u>calories</u>.

_____ 13. One calorie is the amount of energy needed to raise the temperature of pure water <u>10° Fahrenheit</u>.

_____ 14. One <u>million</u> calories is equal to 1 Food calorie.

_____ 15. Transfer of heat energy through a solid is called <u>conduction</u>.

_____16. Transfer of heat energy through a liquid is called <u>convection</u>.

_____17. Transfer of heat energy through a gas is called <u>radiation</u>.

_____18. Transfer of heat energy through a vacuum is called <u>evaporation</u>.

_____19. Energy given off by the rapid motion of electrons is called <u>electromagnetic</u> energy.

_____20. Liquid <u>mercury</u> freezes at 32°F, 0°C and 273K.

PROBLEM

Directions: Read the paragraph and answer questions #21 through #25. *10 points*

An Italian scientist named Galileo Galilei invented one of the first thermometers. His invention was called a *thermoscope* and looked like the apparatus shown in Figure A. The shaded area represents the water that Galileo used to fill his device. The area above the liquid inside the small glass bulb at the top is filled with air. The lower flask is also filled with air and opened to the outside atmosphere.

FIGURE A

21. What would probably happen to the level of the water in the tube if Galileo lit a match by the upper bulb and held it there?

_____

22. What would probably happen to the level of the water in the tube if Galileo lit a match by the lower flask and held it there?

_____

23. What would probably happen to the level of the water in the tube if Galileo rubbed the top bulb with ice?

_____

24. What would probably happen to the level of the water in the tube if Galileo rubbed the lower flask with ice?

_____

25. Would the position of the liquid in the tube tend to be higher or lower in winter than in summer?

_____

_____   _____   ____/____/____
Student's Signature   Parent's Signature   Date

CH5 PHYSICAL AND CHEMICAL CHANGE

Teacher's Classwork Agenda and Content Notes

Classwork Agenda for the Week

1. Students will be able to explain the difference between a physical and a chemical change.
2. Students will separate two simple mixtures by physical means.
3. Students will perform two chemical reactions.
4. Students will be able to distinguish between exothermic and endothermic chemical reactions.

Content Notes for Lecture and Discussion

The study of physics and the study of chemistry were two distinct disciplines at the start of the 19th century. Due largely to the work of **Sir Isaac Newton** (b. 1642; d. 1727), the study of physics was confined primarily to the study of mechanics and the energy associated with moving bodies. Newton's famous contemporary, the German mathematician **Gottfried Wilhelm Leibniz** (b. 1646; d. 1716), recognized the equivalence of potential and kinetic energy; and, the **Law of Conservation of Energy** became the predominating rule of physics. The study of chemistry, on the other hand, examined the changes in bonding structure among the particles that comprised matter when substances were altered in form or physical properties. Ironically, many chemists of the time had not yet accepted the notion that atoms were real entities. Atoms merely served as hypothetical constructs that helped to explain a variety of phenomena. Chemistry was joined to physics with the work of the Swiss-Russian chemist **Germain Hess** (b. 1802; d. 1850).

In 1840, Hess published the **law of constant heat summation: Hess's law**. It was generally known that chemical reactions—such as the burning of wood and coal—involved a transfer of heat. The work of **Joseph Black** (b. 1728; d. 1799) had helped to clarify the idea of **specific heat** which allowed chemists to measure the energy content of reacting materials. Hess discovered that the energy absorbed or released when substances changed from one particular form to another was the same regardless of the route taken in creating the products from the reactants. So, Substance A could be combined with Substance B to form Substance C. And, Substance C could be combined with Substance D to form Substance E. However, the total energy absorbed or released during this sequence of chemical reactions was the same despite the intermediary substances used in producing Substance E from Substances A and B. Hess showed that the pathway one used to produce products from reactants was immaterial to the energy exchanges involved. It was the nature of the substances themselves that determined the degree of energy transfer. This rule was consistent with the Law of Conservation of Energy and convinced physicists that chemistry behaved like the laws of physics. The science of **physical chemistry** was born.

In studying how energy causes matter to change form, chemists distinguish between changes of phase (i.e, **physical change**) and the alteration of a substance's physical properties (i.e., **chemical change**). A physical change merely alters the shape and form of a substance by freeing or restricting the movement of the particles comprising that substance. But a chemical change alters the physical properties of the particles. A **physical property** is any characteristic of a substance that can be measured. The French chemist **Michel-Eugéne Chevreul** (b. 1786; d. 1889) established **melting point** and **boiling point** as essential physical properties of chemical substances. Color, elasticity, durability, ductility, and brittleness are other measurable physical properties. A chemical change involves an alteration in these measurable quantities and suggests a complete rearrangement of the atoms that determine a substance's behavior.

CH5 Content Notes *(cont'd)*

In Lesson #1 students will learn the difference between a physical and a chemical change and will identify examples of each.

In Lesson #2, students will separate two simple mixtures by physical means and carefully identify the physical properties of the substances involved in each separation.

In Lesson #3, students will perform two chemical reactions again paying close attention to the physical properties of the reactants and products in each chemical change.

In Lesson #4, students will learn the difference between exothermic and endothermic chemical reactions. They will perform a simple demonstration to show how energy can be released by the simple dissolution of a substance. In that activity, sodium hydroxide is hydrated with water. Energy is released in the process because the lattice structure of the ionic base, sodium hydroxide, contains more energy than the ionized, hydrated atoms and molecules that go into solution.

ANSWERS TO THE HOMEWORK PROBLEMS

1. Answers will vary.
2. Answers will vary.
3. Answers will vary.

4. The preparation of the meal must have involved more endothermic reactions because the energy content of the meal was greater than the ingredients that went into it.

ANSWERS TO THE END-OF-THE-WEEK REVIEW QUIZ

1. true	6. synthetics	11. reactants	16. P	21. P	26. 150 calories
2. chemists	7. true	12. products	17. P	22. P	27. 100 calories
3. true	8. celluloid	13. oxygen	18. P	23. C	28. 25 calories
4. some	9. true	14. true	19. C	24. C	and, Figure A is
5. bronze	10. true	15. true	20. P	25. P	circled for 2 points

FIGURE A

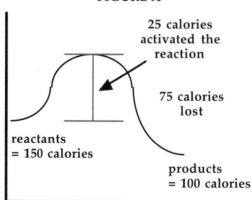

25 calories activated the reaction

75 calories lost

reactants = 150 calories

products = 100 calories

CH5 FACT SHEET

PHYSICAL AND CHEMICAL CHANGE

CLASSWORK AGENDA FOR THE WEEK

(1) Explain the difference between physical and chemical change.
(2) Use a simple physical change to separate a mixture.
(3) Perform a chemical change.
(4) Explain the difference between exothermic and endothermic chemical reactions.

All objects are made of matter and energy can cause matter to change. Some changes occur naturally while other changes are caused by people. All of modern technology is based on our ability to change matter from one form to another. **Chemists** study how these changes takes place and try to discover how different forms and combinations of matter might be used to better our lives.

The word chemistry probably comes from the Greek word *khumos* which means the "juice of a plant." The word *khemeia* refers to the "art of extracting juices from plants." It is certain that even our ancestors realized the value of plant extracts. There is evidence that even Neanderthal Man used plant extracts to treat wounds. In 3,000 B.C. the Mesopotamian cultures of the Middle East discovered how to remove copper and tin from their ores to make bronze. That was the start of the science of **metallurgy**. In modern times, chemists have discovered many ways of making entirely new materials that mother nature did not create. Manmade materials are called **synthetics**. In 1862, the English chemist **Alexander Parkes** (b. 1813; d. 1890) succeeded in manufacturing the first **plastic** from a plant extract called cellulose nitrate. His invention was improved by **John Wesley Hyatt** (b. 1837; d. 1920) in 1869. The plastic that Hyatt "synthesized" is called **celluloid**. Celluloid is the material used to make movie film. In the past century, chemists have created thousands of useful synthetics from **cellophane** and **plexiglass** to **nylon** and **polyvinyl chloride** (i.e., PVC pipe).

Energy can be used to change matter from one shape to another—or from one phase to another—without changing the properties of the matter itself. This kind of change is called a *physical change*. In a **physical change** no new substances are produced. In the unit entitled "The Properties and Phases of Matter," you performed several physical changes by transforming water to ice and water to steam. In each case, the water remained water although it changed from a liquid to a solid or a liquid to a vapor. Water is a distinct substance with specific "chemical properties." When matter changes in size, shape, or phase, that is a physical change.

Energy can also be used to change matter from one kind to another by changing the chemical properties of the materials involved. This type of change is called a chemical change. In a **chemical change** energy interacts with matter to produce entirely new substances. The materials that are mixed together in a chemical "reaction" are called **reactants**. The new substances that are formed are called **products**. Here is a chemical reaction called photosynthesis. Photosynthesis occurs in plants.

$$\text{carbon dioxide gas + water} \quad \rightarrow \quad \text{sugar + oxygen gas}$$
$$\text{"reactants"} \qquad\qquad\qquad \text{"products"}$$

Chemists have always known that chemical reactions involve a transfer of heat energy. In 1840, the Swiss-Russian chemist **Germain Hess** (b. 1802; d. 1850) discovered the **law of constant heat summation**—now called **Hess's Law**. According to Hess's Law, the heat change in a chemical reaction depends on the nature of the reactants and products. The heat change does not depend on the method used to get the reactants to change. The reactants always have a certain amount of energy as do the products. If the reactants have more energy than the products, then energy was released during the chemical reaction. This type of reaction is called an **exothermic reaction**. If the reactants have less energy than the products, then energy must have been absorbed during the chemical reaction. This type of reaction is called an **endothermic reaction**.

Homework Directions

Directions: Write a few sentences that accurately answer questions #1 through #4.

1. Why is dissolving salt in water a physical change?
2. Why is melting (not burning) plastic a physical change?
3. Does baking a batch of chocolate chip cookies produce physical changes, chemical changes, or both? Explain in detail.
4. A chef measured the calorie content of his recipe ingredients before he started cooking and estimated the ingredients to contain 400 Food Calories. After preparing the meal, he found the calorie content of the food to be about 600 Food Calories. Did the preparation of the meal involve mostly exothermic reactions or endothermic reactions? Explain.

Assignment due: _____

_____	_____	____/____/____
Student's Signature	Parent's Signature	Date

CH5 Lesson #1

PHYSICAL AND CHEMICAL CHANGE

Work Date: _____/_____/_____

LESSON OBJECTIVE

Students will be able to explain the difference between a physical and a chemical change.

Classroom Activities

On Your Mark!

Hold up a piece of paper and tear it in half. Ask: Is tearing paper a physical change or a chemical change? Point out that only the size and shape of the paper changed when you tore it but not its properties. Explain that in a physical change no new substances are produced. Instruct students to write the definition of a "physical change" on Journal Sheet #1. A *physical change* is a change in matter in which no new substances are made. Light a match to the paper in a bowl or petri dish and allow it to burn completely to ash. Ask: Is burning paper a physical or a chemical change? Have students identify the substances that were produced (ash, smoke, etc.) as the paper burned. Ask: Does the ash have the same properties as the paper? Light a match to the ash to demonstrate that it does not burn. Conclude that ash is an entirely different substance than paper with entirely different properties (i.e., ash is not combustible). Instruct students to write the definition of a chemical change on Journal Sheet #1. A *chemical change* is a change in matter in which new substances with new properties are made.

Get Set!

Pour a tablespoon of ammonium dichromate powder into a ceramic crucible. Have students record the visible properties of the powder (i.e., solid, orange color, crystalline texture). Bend a two-inch strip of magnesium ribbon at the tip and place it like a flag in the ammonium dichromate powder. Have students record the visible properties of the magnesium ribbon (i.e., solid, grayish/silver color, metallic). Turn out the lights and ignite the ribbon using a Bunsen burner to insure a quick ignition. When the reaction stops, turn on the lights and have students examine the products produced (i.e., a fluffy green powder, a smelly gas, a brittle white crystal buried in the powder). This was definitely a chemical change!

Go!

Instruct students to neatly complete the assignment on Journal Sheet #1 giving a specific explanation for their choice.

Materials

paper, matches, a bowl or petri dish, ceramic crucible, Bunsen burner, ammonium dichromate, magnesium ribbon

CH5 Journal Sheet #1

PHYSICAL AND CHEMICAL CHANGE

Directions: Classify each of the following events as either a physical or a chemical change by writing the word "physical" or "chemical." Then, write a brief sentence to explain your answer.

A. burning paper: _____

_____

B. melting snow: _____

_____

C. sawing wood: _____

_____

D. digesting candy: _____

_____

E. dissolving sugar: _____

_____

F. lighting a candle: _____

_____

G. evaporating puddle: _____

_____

H. bursting balloon: _____

_____

I. crushing rocks: _____

_____

PHYSICAL AND CHEMICAL CHANGE

Work Date: ____/____/____

LESSON OBJECTIVE

Students will separate two simple mixtures by physical means.

Classroom Activities

On Your Mark!

Begin with a review of the definition of a "mixture" as defined in the unit entitled Mixtures. <u>A mixture is any combination of substances that can be separated by physical means</u>. Have students read the definition of a physical change recorded in Lesson #1.

Get Set!

Have students examine and record the visible properties of salt crystals, sand, and water. Have them list any number of ways they could separate the solute from a solvent in a salt solution and a sand solution.

Go!

Assist students in completing the two short activities described in Figure A and Figure B on Journal Sheet #2. Have them closely examine the contents of the crucible and pie pan after the activity described in Figure A. Have them accurately describe the contents of filtered fluid after the activity described in Figure B. Point out that no new materials were produced in these activities. Both involved physical changes: the separation of mixtures by physical means.

Materials

hot plate, pie pan, ceramic crucible or baby food jar, glass funnel, coffee filter or laboratory grade filter paper, beaker, common table salt, sand, water

CH5 JOURNAL SHEET #2

PHYSICAL AND CHEMICAL CHANGE

FIGURE A

<u>Directions</u>: (1) Clean and dry a small ceramic crucible making sure to remove all spots and stains. (2) Place the metal pie pan on a hot plate. (3) Pour 10–15 ml of water into the crucible and stir in 5–10 grams of common table salt for 30 seconds. (4) Place the crucible in the center of the pie pan. (5) Position an inverted glass funnel over the crucible as shown. (6) Turn on the hot plate to a medium setting. (7) Allow the water sufficient time to boil out of the crucible and collect in the pie pan. (8) Turn off the hot plate and let the set-up cool. (9) Examine the contents of the crucible and the pie pan. (10) Was this a physical or a chemical change? Explain your answer.

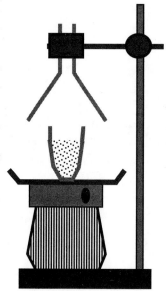

GENERAL SAFETY PRECAUTIONS

Be sure you are familiar with the proper use of the hot plate. Wear goggles to protect your skin and eyes from being burned by HOT WATER. Do not touch any part of the equipment without heat-resistant gloves or tongs. Clean up only when the apparatus is cool.

FIGURE B

<u>Directions</u>: (1) Mix 100 ml of water and 20 grams of sand into a small beaker. (2) Fold a coffee or laboratory filter to fit inside a glass funnel. (3) Place the funnel in an Ehrlenmeyer flask. (4) Slowly pour the muddy mixture through the paper filter to separate the mixture. (5) Was this a physical or a chemical change? Explain your answer.

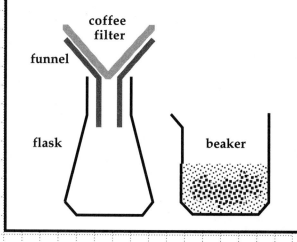

CH5 Lesson #3

PHYSICAL AND CHEMICAL CHANGE

Work Date: ____/____/____

LESSON OBJECTIVE

Students will perform two chemical reactions.

Classroom Activities

On Your Mark!

Review the definition of a "chemical change." <u>A chemical change is a change in matter in which new substances with new properties are made.</u> Have students read the definition of a chemical change recorded in Lesson #1.

Get Set!

Have students examine and record the visible properties of Epsom salt and borax crystals. Remind them that Epsom salt (chemical name: magnesium sulfate) is an orthorhombic crystal. Inform them that Borax detergent contains hydrated sodium borite consisting of monoclinic crystals. Explain that household ammonia is a water solution of ammonium hydroxide: a chemical "base" similar to that used in most soaps. Elmer's white glue is a plant extract made of cellulose which is rich in the elements carbon and hydrogen. **Be sure students are aware of the toxicity of these substances. They can be extremely caustic.**

Go!

Assist students in completing the two short activities described in Figure C and Figure D on Journal Sheet #3. Have them closely examine the products of these activities and distinguish them from those of the reactants. Review the terms **reactants** and **products** as defined in the Fact Sheet. Point out that the new materials formed in the first activity were ammonium sulfate and magnesium hydroxide which is commonly called "milk of magnesia." Milk of magnesia is used as an antacid and a laxative. The second activity involved a **polymerization** reaction. In this polymerization reaction, the sodium borite caused the carbon atoms in the glue to become linked together in complex chains. The network of chains gives this new product its "rubbery" properties. Both activities involved chemical changes.

Materials

protective gloves, paper cups, Ehrlenmeyer flasks, glass stirring rods, glass funnel, coffee filter or laboratory grade filter paper, beakers, small graduated cylinders, Borax detergent, household ammonia cleaner, Epsom salt, Elmer's glue

CH5 JOURNAL SHEET #3

PHYSICAL AND CHEMICAL CHANGE

FIGURE C

General Safety Precaution: Do not inhale ammonia fumes. Ammonia can burn the eyes and the skin. Wear goggles, apron, and protective gloves when handling ammonia.

Directions: (1) Measure out 10 grams of Epsom salt. (2) Briefly examine some of the crystals under a microscope or hand-held magnifying glass. You may notice that the crystals are orthorhombic crystals. (3) Fill a small beaker with 100 ml of water. (4) Stir the Epsom salt into the water. (5) Pour 5 ml of household ammonia into a small graduated cylinder. (6) Add the liquid ammonia to the Epsom salt mixture but DO NOT STIR. (7) Allow the solution to stand for ten minutes. (8) Gently pour the "milky white" mixture through a paper filter as you did in Lesson #2 with the sand mixture. (9) Allow the substance in the filter to dry for 24 hours and examine the contents under the microscope. (10) Would you classify these crystals in the same class as Epsom salt crystals? Was this a physical or a chemical change?

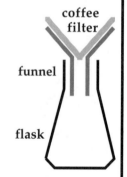

epsom salt mixture

household ammonia

coffee filter

funnel

flask

FIGURE D

Directions: (1) Measure out 10 grams of Borax detergent. (2) Examine and describe the properties of the crystals (color, size, etc.). (3) Pour the detergent into a paper cup. (4) Slowly add several drops of Elmer's white glue and stir using a wood stirring rod. (5) Continue to add the glue until the mixture becomes moist and rubbery. (6) Remove the mixture and roll it around in your hands until it forms a small grayish ball. (7) How does this substance compare with the substances you used to make it? Was this a physical or a chemical change?

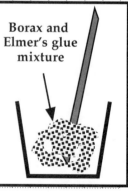

Borax and Elmer's glue mixture

PHYSICAL AND CHEMICAL CHANGE

Work Date: ____/____/____

LESSON OBJECTIVE

Students will be able to distinguish between exothermic and endothermic chemical reactions.

Classroom Activities

On Your Mark!

Begin with a review of the chemical reactions performed in Lesson #1 and Lesson #3. Point out that all of these chemical reactions involved an exchange of heat. In any chemical reaction, the energy stored in the molecular reactants is almost always different than the energy stored in the molecular products. Review the concept of **specific heat** as defined in Lesson #2 of the unit entitled Heat and Energy Transfer. The specific heat of a substance is the amount of energy needed to raise the temperature of that substance 1°C. For example: The specific heat of water is 1.0. The specific heat of aluminum is 0.22. Energy can either be released or absorbed by a chemical reaction. If the energy of the products is greater than the energy of the reactants, then energy must be added to the reactants to produce the products. This type of reaction is an **endothermic reaction**. If the energy of the products is less than the energy of the reactants, then energy is given off by the reactants as they form the products.

Get Set!

Draw Illustration A on the board and have students fill out the diagrams shown on Journal Sheet #4 to illustrate what happens to the energy of the reactants and products during a chemical reaction. Point out that even though the reactants might have more energy than the products, it is sometimes necessary to "kick off" the reaction by adding **activation energy** (e.g., a match).

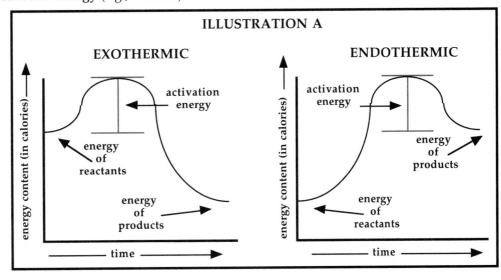

ILLUSTRATION A

EXOTHERMIC — energy content (in calories) — activation energy — energy of reactants — energy of products — time

ENDOTHERMIC — energy content (in calories) — activation energy — energy of products — energy of reactants — time

Go!

Assist students in performing the activity described in Figure E of Journal Sheet #4.

Materials

ring stand and clamps, thermometers, glass stirring rods, sodium hydroxide pellets, water, beakers

CH5 JOURNAL SHEET #4

PHYSICAL AND CHEMICAL CHANGE

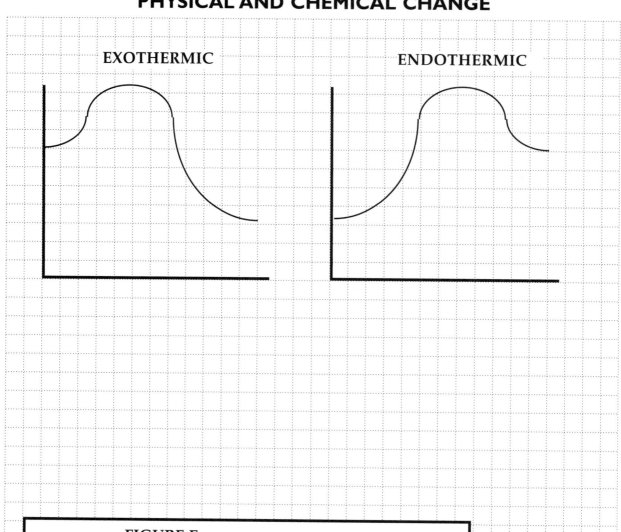

EXOTHERMIC   ENDOTHERMIC

FIGURE E

Directions: (1) Measure out 10 grams of solid sodium hydroxide pellets (i.e., lye) onto a small piece of paper. (2) Fill a 250 ml beaker with 50-100 ml of water. (3) Secure a thermometer to a ring stand and lower it into the beaker as shown. (4) Wait two minutes before recording the temperature of the water. (5) Slowly stir the sodium hydroxide pellets into the beaker using a glass stirring rod. Make sure not to hit the thermometer. (6) Record the temperature of the water every 15 seconds for the next 5 minutes. (7) Did the dissolving of the sodium hydroxide produce an endothermic or exothermic reaction?

GENERAL SAFETY PRECAUTIONS

Wear goggles and an apron. Sodium hydroxide is a caustic base that can burn the skin and eyes. Wipe up all drops and spills immediately.

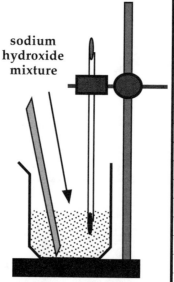

sodium hydroxide mixture

CH5 REVIEW QUIZ

Directions: Keep your eyes on your own work.
Read all directions and questions carefully.
THINK BEFORE YOU ANSWER!
Watch your spelling, be neat, and do the best you can.

CLASSWORK   (~40): _____
HOMEWORK   (~20): _____
CURRENT EVENT   (~10): _____
TEST   (~30): _____

TOTAL   (~100): _____
(A ≥ 90, B ≥ 80, C ≥ 70, D ≥ 60, F < 60)

LETTER GRADE: _____

TEACHER'S COMMENTS: _____

_____

PHYSICAL AND CHEMICAL CHANGE

TRUE–FALSE FILL-IN: If the statement is true, write the word TRUE. If the statement is false, change the underlined word to make the statement true. *15 points*

_____ 1. Energy <u>can</u> cause matter to change.

_____ 2. <u>Biologists</u> study how changes take place in matter and try to discover new combinations of matter.

_____ 3. The word chemistry probably comes from the Greek word *khumos* which means the "<u>juice of a plant</u>."

_____ 4. There is <u>no</u> evidence that Neanderthal Man used plant extracts to treat wounds.

_____ 5. Copper and tin are combined to make <u>steel</u>.

_____ 6. Manmade materials are called <u>synonyms</u>.

_____ 7. In 1862, Alexander Parkes succeeded in manufacturing <u>plastic</u>.

_____ 8. In 1869, John Wesley Hyatt synthesized <u>nylon</u>.

_____ 9. In a <u>physical</u> change no new substances are produced.

_____ 10. In a <u>chemical</u> change energy interacts with matter to produce entirely new substances.

_____ 11. The materials that are mixed together in a chemical reaction are called <u>products</u>.

_____ 12. The materials formed in a chemical reaction are called <u>reactants</u>.

_____ 13. In photosynthesis, carbon dioxide gas and water produce sugar and <u>hydrogen</u> gas.

_____ 14. A chemical reaction that absorbs energy is an <u>endothermic</u> reaction.

_____ 15. A chemical reaction that releases energy is called an <u>exothermic</u> reaction.

Directions: In the space provided, write the letter "P" if the described change is a physical change and the letter "C" if it is a chemical change. *10 points*

_____ 16. breaking a coffee cup

_____ 17. tearing cellophane

_____ 18. inflating a balloon

_____ 19. burning a log

_____ 20. boiling water

_____ 21. dissolving sugar

_____ 22. evaporating liquid mercury

_____ 23. digesting a hamburger

_____ 24. rusting a nail

_____ 25. slicing a steak

Directions: Read the paragraph below and answer questions #26 through #28, then circle Figure A or Figure B to show which graph helped you to answer these questions. *5 points*

A scientist measured the calorie content of a given amount of Substance A and found it to be 50 calories. He measured the calorie content of a given amount of Substance B and found it to be 100 calories. He mixed Substance A and Substance B together and nothing happened. He put the mixture onto a hot plate and added heat. During the chemical reaction 75 calories of energy was released. After the reaction was finished, the scientist measured the calorie content of the newly formed substances. The calorie content of newly formed Substance C was 60 calories. The calorie content of newly formed substance D was 40 calories.

26. What was the total energy content of the reactants? _____

27. What was the total energy content of the products? _____

28. How much energy was needed to "activate" the reaction? _____

FIGURE A

FIGURE B

_____
Student's Signature

_____
Parent's Signature

____/____/____
Date

ELEMENTS, MOLECULES, AND COMPOUNDS

TEACHER'S CLASSWORK AGENDA AND CONTENT NOTES

Classwork Agenda for the Week

1. Students will use chemical symbols and formulas to illustrate the difference between elements, molecules, and compounds.

2. Students will build models to visualize the difference between elements, molecules, and compounds.

3. Students will demonstrate that water is a compound.

4. Students will write and balance chemical equations to show how matter is conserved in a chemical change.

Content Notes for Lecture and Discussion

The Greek philosopher **Empedocles** (b. 493 B.C.; d. 433 B.C.) was the first to propose that the universe was composed of four elements: earth, air, fire, and water. He believed that the elements were assembled, disassembled, and reassembled by the forces of "harmony" and "disharmony" (i.e., love and discord). His younger contemporary **Democritus** (b. 460 B.C.; d. 370 B.C.) proposed the first **atomic theory** of the universe which suggested that matter was composed of indivisible particles which he called atoms. The Greek word *atom* means "cannot be broken." **Aristotle** (b. 384 B.C.; d. 322 B.C.) was primarily responsible for popularizing the works of Empedocles and Democritus and for laying the foundation for the systematic study of the elements.

In 1661, **Robert Boyle** (b. 1627; d. 1691)—in his publication *Sceptical Chymist*—gave us the modern definition of an element and argued that there must be considerably more than four. According to Boyle, an element is a simple immutable substance that can neither be decomposed nor transformed or composed from other chemical substances. In short, an **element** is a pure substance that cannot be broken down into simpler form by ordinary chemical means. The work of 18th Century chemists resulted in the discoveries of as many as 30 elements which the French chemist **Antoine Laurent Lavoisier** (b. 1743; d. 1794) listed in his 1789 work *Traité élémentaire de chimie*. Lavoisier himelf was responsible for proving that the "element of fire" was imaginary by demonstrating that combustion required only a portion of the air. He also suggested a method of naming chemical compounds according to the elements they contain. With the French astronomer and mathematician **Pierre Simon Laplace** (b. 1749; d. 1827) Lavoisier showed in 1783 that water was a compound composed of the elements hydrogen and oxygen. In 1766, the English chemist **Henry Cavendish** (b. 1731; d. 1810) discovered hydrogen, or "factitious air," by reacting acids with metals and alkalis. In 1774, **Joseph Priestley** (b. 1733; d. 1804), an English chemist and Unitarian minister, discovered oxygen by heating mercuric oxide with focused sunlight. Priestley also perfected a way of pressurizing carbon dioxide gas and dissolving it in water. His method allowed a group of enterprising businessmen to market the first artificially carbonated beverage which became the rage of Europe. By 1869, the Russian chemist **Dmitri Mendeleev** (b. 1834; d. 1907) was able to arrange more than 60 elements according to their weights and chemical reactivity. His was the first *Periodic Table of the Elements*.

The French physicist and philosopher **Pierre Gassendi** (b. 1592; d. 1655) used the term "molecule" in his book on the theory of atoms, *Syntagma Philosophicum*. Later chemists adopted the term to refer to conglomerations of atoms held together by "shape" or "affinity"—aggregates capable of a "separate existence." The work of Italian physicist **Amedeo Avogadro** (b. 1776; d. 1856) allowed scientists to calculate the number of atoms or molecules present in a given volume of gas. **Avogadro's number** (6.023×10^{23} atoms or molecules) represents the number of atoms or molecules present in **one mole** of a substance. Avogadro's number made it possible for chemists to accurately determine the atomic weights of atoms, molecules, and compounds.

Prior to the 19th century, a compound was considered nothing more than a closely fixed mixture and the distinction between compounds and mixtures was vague. Many chemists expressed the notion that compounds could change in composition. The chemists **Joseph Louis Proust** (French–b. 1754; d. 1826) and **John Dalton** (English–b. 1766; d. 1844) laid that idea to rest. They insisted that compounds must be considered as constant substances comprised of definite proportions of elements. According to Proust and Dalton, a compound is a substance composed of chemically combined elements existing in constant proportion.

Chemical symbols, **chemical formulas**, and **chemical equations** are the "letters," "words," and "sentences" that form the syntax of the language of chemistry. The **Law of Conservation of Matter and Energy** is chemistry's primary grammatical rule. According to the Law of Conservation of Matter and Energy matter can neither be created nor destroyed. The study of chemistry is the study of how particles of matter rearrange themselves to give the universe its diversity of material substances.

ANSWERS TO THE HOMEWORK PROBLEMS

1. Be sure that students copy the *Periodic Table* accurately and obey the rules for writing chemical symbols. The first letter of the symbol is *always* capitalized; a second is *always* lowercase: sodium—Na; calcium—Ca; carbon—C; nitrogen—N; oxygen—O; chlorine—Cl; phosphorus—P; silicon—Si; magnesium—Mg (as opposed to manganese—Mn); krypton—Kr

2. sodium chloride—NaCl (1 atom of sodium, 1 atom of chlorine)
   sulfuric acid—H_2SO_4 (2 atoms of hydrogen, 1 atom of sulfur, 4 atoms of oxygen)
   potassium chlorate—$KClO_3$ (1 atom of potassium, 1 atom of chlorine, 3 atoms of oxygen)
   calcium carbonate—$CaCO_3$ (1 atom of calcium, 1 atom of carbon, 3 atoms of oxygen)
   hydrogen sulfide—H_2S (2 atoms of hydrogen, 1 atom of sulfur)

3. Balance this chemical equation: $Zn + 2HCl \rightarrow H_2 + ZnCl_2$

ANSWERS TO THE END-OF-THE-WEEK REVIEW QUIZ

1. true	6. elements	11. true	16. 24
2. atom	7. true	12. pairs	17. not/never
3. true	8. true	13. O_2	18. not/never
4. element	9. always	14. true	19. photosynthesis
5. atom	10. always	15. true	20. coefficients

(A) $2H_2 + O_2 \rightarrow 2H_2O$

(B) $2KClO_3 \rightarrow 2KCl + 3O_2$

(C) $2Na + 2H_2O \rightarrow H_2 + 2NaOH$

(D) $NaOH + HCl \rightarrow NaCl + H_2O$ (already balanced)

(E) $2HgO \rightarrow 2Hg + O_2$

CH6 FACT SHEET

ELEMENTS, MOLECULES, AND COMPOUNDS

CLASSWORK AGENDA FOR THE WEEK

(1)   Use chemical symbols and formulas to illustrate the difference between elements, molecules, and compounds.
(2)   Build models to visualize the difference between elements, molecules, and compounds.
(3)   Demonstrate that water is a compound.
(4)   Write and balance chemical equations to show how matter is conserved in a chemical change.

The early Greeks believed there were four basic particles that made up all matter. According to the Greeks, these four basic **elements** were made of **earth** particles, **air** particles, **water** particles, and **fire** particles. They thought that every substance they could see, smell, taste, or feel was made of these four types of particles. A Greek philosopher named **Democritus** (b. 460 B.C.; d. 370 B.C.) called the smallest particle of an element an **atom**. Today we know there are a lot more than four elements (i.e., four kinds of atoms). We also know that atoms are not the most basic or simplest units of matter.

The definition of an element is much the same today as it was long ago. An **element** is any substance that cannot be split into simpler substances by ordinary chemical means. Twelve grams of pure carbon, for example, is composed of about six hundred billion trillion (i.e., 6×10^{23}) separate atoms of carbon; and, every atom of carbon in that sample behaves exactly like every other atom of carbon when mixed with other substances in a chemical reaction. So, an **atom** is defined as the smallest part of a chemical element.

The chart that shows all of the known chemical elements is called **The Periodic Table of the Elements**. There are more than 100 elements on that chart. Each element is represented by one or two letters called a **chemical symbol**. The English or Latin name of a chemical element is used to decide which letters will represent that element. For example, the element **hydrogen**—which makes up about 90 percent of the universe—is represented by the capital letter "**H**". **Iron** is represented by the letters "**Fe**" from the Latin word for iron: ferrum. The letter "**I**" is used to represent the element **iodine**. If an element is represented by one letter, then that letter is always capitalized. If the element is represented by two letters, then the first letter is always capitalized and the second letter is always lowercase. Chemists obey these simple rules for writing chemical symbols to avoid mistakes when mixing chemicals together during experimentation.

A **molecule** is a particle that has two or more atoms bonded together. The air we breathe contains molecules of oxygen atoms bonded together in pairs. A chemist will write the chemical symbol for oxygen as "**O**". However the **chemical formula** for the oxygen molecules that we breathe is written "**O_2**". The little "2" at the lower right corner of the symbol for oxygen tells us that there are *two oxygen atoms in one molecule of oxygen gas*. The nitrogen gas we breathe is also made of molecules. Nitrogen gas contains the substance N_2: two atoms of nitrogen bonded together to form one molecule of nitrogen gas.

A **compound** is a molecule that has more than one kind of atom bonded together. It is a substance composed of two or more chemically combined elements. *Water is made of two hydrogen atoms bonded to one oxygen atom.* A chemist writes the formula for water as "**H_2O**." It is not necessary to write a small number "1" next to the symbol for oxygen because the symbol itself stands for one atom of oxygen. The chemical formula for a simple sugar called **glucose** is $C_6H_{12}O_6$. Every glucose molecule contains 6 atoms of carbon, 12 atoms of hydrogen, and 6 atoms of oxygen bonded together: 24 atoms in all to make one glucose molecule.

Scientists believe that no new matter or energy is ever created anew or destroyed completely. This principle is called the **Law of Conservation of Matter and Energy**. All chemical reactions obey this law. This means that in any chemical reaction, no new atoms are ever added or taken away from the reaction. The atoms are merely rearranged to form new substances. Plants carry on **photosynthesis** which uses the energy of the sun to make glucose and oxygen out of carbon dioxide and water. A chemist writes the chemical equation to describe this process as follows:

$$6CO_2 \quad + \quad 6H_2O \quad \rightarrow \quad C_6H_{12}O_6 \quad + \quad 6O_2$$

"carbon dioxide gas" and "water" form "glucose" and "oxygen gas"

The large number "6" in front of the chemical formulas for "carbon dioxide gas," "water," and "oxygen gas" are called **coefficients**. These coefficients tell us that *six* molecules of each of these substances was involved in the production of *one* molecule of "glucose." Note: There is no coefficient in front of the chemical formula for glucose because the formula itself represents one molecule of glucose. Coefficients are used to "balance" chemical equations in order to satisfy the Law of Conservation of Matter and Energy.

Homework Directions

1. Use The Periodic Table of the Elements given to you by your instructor to write the chemical symbols for the following elements: sodium, calcium, carbon, nitrogen, oxygen, chlorine, phosphorus, silicon, magnesium, krypton.

2. Write the number of atoms for each element making up each of the following compounds:

    sodium chloride—NaCl   sulfuric acid—H_2SO_4 potassium chlorate—$KClO_3$

    calcium carbonate—$CaCO_3$ hydrogen sulfide—H_2S

3. Balance this chemical equation: ___ Zn + ___ HCl → ___ H_2 + ___ $ZnCl_2$

Assignment due: _____

_____   _____   ___/___/___
Student's Signature   Parent's Signature   Date

ELEMENTS, MOLECULES, AND COMPOUNDS

Work Date: ___/___/___

LESSON OBJECTIVE

Students will use chemical symbols and formulas to illustrate the difference between elements, molecules, and compounds.

Classroom Activities

On Your Mark!

If you have access to the compound mercuric oxide (HgO) perform the simple demonstration shown in Illustration A to demonstrate the dissociation of a compound into its constituent elements.

Display samples of common <u>elements</u> (e.g., bare copper wire, zinc chips, charcoal), an empty covered jar to represent diatomic <u>molecules</u> of oxygen gas and nitrogen gas (which comprise 99% of the air we breathe), and a beaker of the <u>compound</u> water. Referring to the Teacher's Agenda and Content Notes, give a brief lecture describing the old theory of elements proposed by **Empedocles** (b. 493 B.C.; d. 433 B.C.), **Democritus** (b. 460 B.C.; d. 370 B.C.), and **Aristotle** (b. 384 B.C.; d. 322 B.C.). Explain that today scientists are aware of more than 100 elements. Have students copy the definitions below onto Journal Sheet #1:

- An **element** is any substance that cannot be split into simpler substances by ordinary chemical means.
- A **molecule** is a particle that has two or more atoms bonded together.
- A **compound** is a molecule with different kinds of atoms.

Get Set!

Distribute copies of the Periodic Table of the Elements appearing in the Appendix or use copies from your district's adopted textbook. Explain the rules for writing chemical symbols and formulas to students as they appear in the Fact Sheet.

Go!

Assist students in writing the chemical symbols of elements and the chemical formulas for molecules and compounds like the examples on Journal Sheet #1. For additional examples, refer to the names and chemical formulas of the other substances used in this volume.

ILLUSTRATION A

<u>Directions</u>: (1) Spill 10–15 grams of mercuric oxide (HgO) into a test tube. (2) Insert a 5 cm length of glass tubing into a one-holed rubber stopper and connect the glass tube to an adequate length of rubber tubing as shown. (3) Fill a second test tube and a beaker with water, invert the test tube into the beaker, and secure. (4) Insert the rubber tubing into the water-filled test tube. (5) Turn on the Bunsen burner and within seconds oxygen gas will collect in the test tube and displace the water. **Joseph Priestley** (b. 1733; d. 1804) used a slower method to discover oxygen in 1774. He used a magnifying glass to focus sunlight onto the HgO-filled test tube. The German chemist **Robert Wilhelm Bunsen** (b. 1811; d. 1899) did not invent the Bunsen burner until 1855.

GENERAL SAFETY PRECAUTIONS

Mercuric oxide is highly volatile. Many school districts no longer allow its use because the reaction produces toxic liquid mercury. DO NOT ALLOW STUDENTS TO HANDLE THIS SUBSTANCE. DO NOT OPEN THE TEST TUBE IN THEIR PRESENCE. STUDENTS OBSERVING SHOULD WEAR GOGGLES AND APRONS, IN CASE OF TUBE BREAKAGE

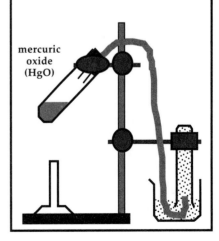

mercuric oxide (HgO)

Materials

materials shown in Illustration A, The Periodic Table of the Elements

CH6 Journal Sheet #1

ELEMENTS, MOLECULES, AND COMPOUNDS

PRACTICE EXAMPLES

Write the chemical formulas for the following compounds:

(A) sodium hydroxide has
    1 atom of sodium
    1 atom of oxygen
    1 atom of hydrogen

(B) silicon dioxide has
    1 atom of silicon
    2 atoms of oxygen

(C) pentane has
    5 atoms of carbon
    12 atoms of hydrogen

ELEMENTS, MOLECULES, AND COMPOUNDS

Work Date: ____/____/____

LESSON OBJECTIVE

Students will build models to visualize the difference between elements, molecules, and compounds.

Classroom Activities

On Your Mark!

Review the distinction between elements, molecules, and compounds. Remind students of the proper way to write chemical symbols and formulas.

Get Set!

Distribute the materials for building molecular models. Decide on the color coding for the individual atoms of each element. Write the color code on the board and have students copy the color code on Journal Sheet #2.

Go!

Assist students in constructing molecular models from the written formulas for each molecule or compound as shown in the examples on Journal Sheet #2. Instruct students to draw a diagram of each molecule that they build (like those in the examples). For additional examples, refer to the names and chemical formulas of the other substances used in this volume.

Materials

wood or plastic models of atoms (if available in commercially sold kits), or toothpicks and 4–5 different colors of clay, or M&M™s and small paper cups

CH6 JOURNAL SHEET #2

ELEMENTS, MOLECULES, AND COMPOUNDS

PRACTICE EXAMPLES

Construct models and write the chemical formulas for the following compounds:

(A) methane has
    1 atom of carbon
    4 atoms of hydrogen

(B) hydrogen peroxide has
    2 atoms of hydrogen
    2 atoms of oxygen

(C) ammonia has
    1 atom of nitrogen
    3 atoms of hydrogen

 hydrogen   ● carbon   ● oxygen   nitrogen

ELEMENTS, MOLECULES, AND COMPOUNDS

Work Date: ____/____/____

LESSON OBJECTIVE

Students will demonstrate that water is a compound.

Classroom Activities

On Your Mark!

Begin with a brief discussion of what early scientists thought of compounds. Explain that prior to the 19th century, a compound was considered nothing more than a closely fixed mixture and the distinction between compounds and mixtures was vague. Mention that many chemists expressed the notion that compounds could change in composition. Have them copy the names of chemists **Joseph Louis Proust** (French–b.1754; d.1826) and **John Dalton** (English–b. 1766; d. 1844) on Journal Sheet #3. According to Proust and Dalton, a compound is a substance composed of chemically combined elements existing in <u>constant proportion</u>. Have them copy the following examples on the board to show them how the atoms of each element in a chemical formula can be expressed as a ratio:

compound name:	water	sulfuric acid	glucose
chemical formula:	H_2O	H_2SO_4	$C_6H_{12}O_6$
ratio of elements:	2:1	2:1:4	6:12:6 reduces to 1:2:1

After separating the elements that made up a given compound, scientists measured the volumes or weights of the elements and came up with ratios like the ones shown above. These ratios gave them a way to determine the chemical formulas for the compounds under study.

Get Set!

Point out that the French chemists **Pierre Simon Laplace** (b. 1749; d. 1827) **Antoine Laurent Lavoisier** (b. 1743; d. 1794) showed in 1783 that water was a compound composed of the elements hydrogen and oxygen. They accomplished this by passing an electrical current through water. Twice as much gas was formed at the negative electrode as was formed at the positive electrode. Thus, there were two invisible gases, each of which preferred a <u>different electrode</u>. They named one gas "hydrogen" meaning "generated from water." They discovered that the other gas was the same gas discovered by Joseph Priestley in 1774—oxygen.

Go!

Prepare a dilute solution of sulfuric acid to act as a catalyst in the experiment described in Figure A on Journal Sheet #3. *Warn students about exercising safety precautions when dealing with an acid and have an eye wash handy in case of accidents.* Assist students in performing the experiment described on Journal Sheet #3. Oxygen is highly soluble in water; therefore, the amount of oxygen captured in this experiment might be less than half the amount of hydrogen. Hydrogen is explosive; so, students will hear an audible "pop" as they ignite the gas in Step #11. Oxygen will not "pop."

Materials

test tubes, 100 ml or 250 ml beakers, insulated copper wire, water, dilute sulfuric acid, D-cell batteries, switches, alligator clips, tape or battery holders

CH6 JOURNAL SHEET #3

ELEMENTS, MOLECULES, AND COMPOUNDS

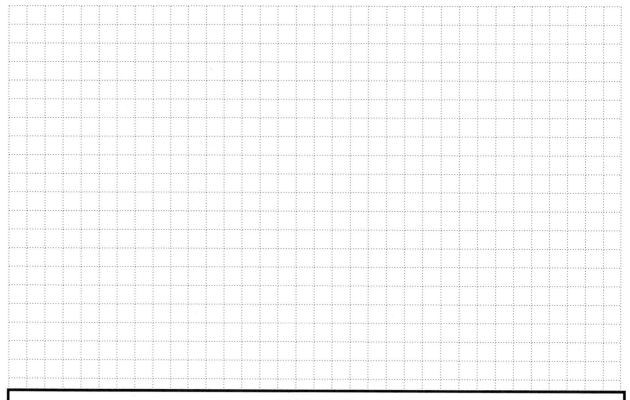

FIGURE A

<u>Directions</u>: (1) Fill a 100 ml beaker with water. (2) Add a medicine dropper full of the acid solution given to you by your instructor. AVOID SPLATTERING THE ACID. (3) Fill two small test tubes with water and without allowing air to enter the tubes invert them into the beaker. (4) Construct the set-up shown using insulated copper wires, alligator clips, two D-cell batteries, and a switch. Be sure the ends of the "electrode wires" are stripped bare. (5) Insert the bare electrodes into the inverted test tubes. (6) Close the switch and record your observations. (7) Allow the gases to accumulate in the test tubes for 10 minutes or until the test tube over the negative electrode is full of gas. (8) Open the switch to shut off the current. (9) Carefully lift the gas-filled test tube over the negative (-) electrode out of the water and keeping it in an inverted position quickly cover the opening with your thumb. (10) Have your assistant light a match. (11) Remove your thumb from the test tube at the same instant your assistant puts the match flame to the opening. Record your observations. (12) Repeat steps #9 through #11 using the test tube over the positive (+) electrode. (13) Explain why this experiment shows that water is a compound.

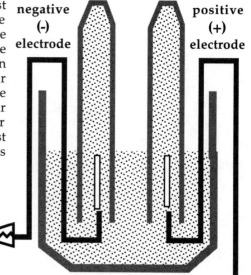

negative
(-)
electrode

positive
(+)
electrode

CH6 Lesson #4

ELEMENTS, MOLECULES, AND COMPOUNDS

Work Date: ____/____/____

LESSON OBJECTIVE

Students will write and balance chemical equations to show how matter is conserved in a chemical change.

Classroom Activities

On Your Mark!

Have students refer to the Fact Sheet and the **chemical equation** that describes the process of **photosynthesis**. Discuss the paragraph and equation thoroughly so that students understand the meaning of each of the symbols, formulas, **subscripts** (i.e., numbers showing the number of atoms in each molecule) and **coefficients** (i.e., numbers showing how many molecules of each substance are involved in the reaction).

Get Set!

Write a series of chemical equations on the board. Refer to the equations used elsewhere in this volume. Do not include the coefficients. Write them with blanks in the place of the coefficients as in Problems "A" through "E" on the End-of-the-Week Quiz. Explain that **chemical symbols**, **chemical formulas**, and **chemical equations** are the "letters," "words," and "sentences" that form the syntax of the language of chemistry. The **Law of Conservation of Matter and Energy** is chemistry's primary rule of grammar. According to the Law of Conservation of Matter and Energy matter can neither be created nor destroyed. In all chemical reactions particles of matter are merely rearranged to give the universe its diversity of material substances.

Go!

Have students refer to the example on Journal Sheet #4. Advise them to use the same sequence (i.e., formulas followed by drawn pictures) to help them derive the coefficients in front of each molecule in the reaction. They may also use molecular models (in lieu of the drawings) to accomplish this. To balance the equations, additional atoms can be added to one side of the equation or the other <u>only</u> if the whole molecule that the atom came from is added as well.

Materials

Journal Sheet #4, toothpicks and clay (or model kits) used in Lesson # 2

CH6 JOURNAL SHEET #4

ELEMENTS, MOLECULES, AND COMPOUNDS

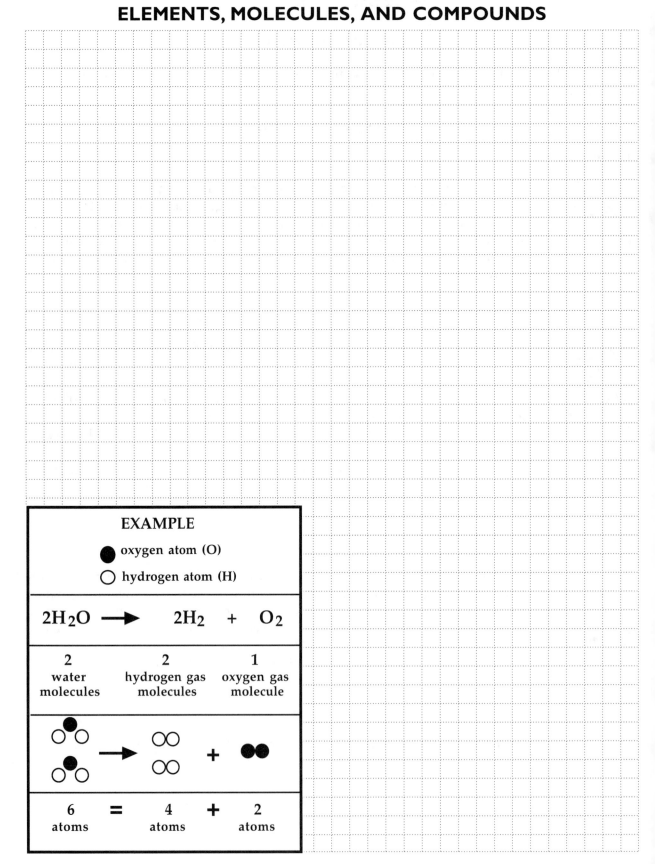

CH6 REVIEW QUIZ

Directions: Keep your eyes on your own work.
Read all directions and questions carefully.
THINK BEFORE YOU ANSWER!
Watch your spelling, be neat, and do the best you can.

CLASSWORK (~40): _____
HOMEWORK (~20): _____
CURRENT EVENT (~10): _____
TEST (~30): _____

TOTAL (~100): _____
(A ≥ 90, B ≥ 80, C ≥ 70, D ≥ 60, F < 60)

LETTER GRADE: _____

TEACHER'S COMMENTS: _____
_____

ELEMENTS, MOLECULES, AND COMPOUNDS

TRUE–FALSE FILL-IN: If the statement is true, write the word TRUE. If the statement is false, change the underlined word to make the statement true. *20 points*

_____ 1. The early Greeks believed there were <u>four</u> basic particles that made up all matter.

_____ 2. A Greek philosopher named Democritus called the smallest particle of an element a(n) <u>miniparticle</u>.

_____ 3. Atoms <u>are not</u> the most basic or simplest units of matter.

_____ 4. A(n) <u>compound</u> is any substance that cannot be split into simpler substances by ordinary chemical means.

_____ 5. A(n) <u>molecule</u> is defined as the smallest part of a chemical element.

_____ 6. The chart that shows all of the known chemical elements is called The Periodic Table of the <u>Chemicals</u>.

_____ 7. There are more than <u>100</u> known elements.

_____ 8. Each element is represented by one or two letters called a chemical <u>symbol</u>.

_____ 9. If an element is represented by one letter, then that letter is <u>sometimes</u> capitalized.

_____ 10. If an element is represented by two letters, then the first letter is always capitalized and the second letter is <u>sometimes</u> lowercase.

_____ 11. A <u>molecule</u> is a particle that has two or more atoms bonded together.

_____ 12. The air we breathe contains molecules of oxygen atoms bonded together in <u>threes</u>.

_____ 13. The chemical formula for the oxygen molecules we breathe is written as follows: <u>O_3</u>.

_____ 14. A <u>compound</u> is a substance composed of two or more chemically combined elements.

_____ 15. A chemist writes the formula for water as follows: <u>H_2O</u>.

_____16. The chemical formula for a simple sugar called glucose is $C_6H_{12}O_6$. This molecule contains <u>36</u> atoms in all.

_____17. Scientists believe that existing matter or energy is <u>sometimes</u> created anew or destroyed completely.

_____18. In all chemical reactions atoms are <u>sometimes</u> destroyed to form new substances.

_____19. Plants carry on <u>respiration</u> which uses the energy of the sun to make glucose and oxygen out of carbon dioxide and water.

_____20. Numbers called <u>subscripts</u> are used to show how many molecules of a substance are involved in a chemical reaction.

PROBLEMS

Directions: Fill in the blanks to balance these chemical equations. *10 points*

(A) ___ H_2 + ___ O_2 → ___ H_2O

(B) ___ $KClO_3$ → ___ KCl + ___ O_2

(C) ___ Na + ___ H_2O → ___ H_2 + ___ $NaOH$

(D) ___ $NaOH$ + ___ HCl → ___ $NaCl$ + ___ H_2O

(E) ___ HgO → ___ Hg + ___ O_2

_____   _____   ____/____/____
Student's Signature   Parent's Signature   Date

ATOMIC STRUCTURE

TEACHER'S CLASSWORK AGENDA AND CONTENT NOTES

Classwork Agenda for the Week

1. Students will draw diagrams of atoms to show how models of atoms have evolved.

2. Students will construct models of Bohr atoms.

3. Students will diagram how atoms can be transformed into ions.

4. Students will diagram how atoms can form ionic or covalent bonds with other atoms.

Content Notes for Lecture and Discussion

The model of the atom has evolved since ancient times and is still evolving today. **Democritus** (b. 460 B.C.; d. 370 B.C.) believed the atom to be indestructible but the study of electrolysis by the English physicist **Michael Faraday** (b. 1791; d. 1867) made it clear that atomic particles could carry electrical charge. Another English physicist, **J. J. Thomson** (b. 1856; d. 1940), used high energy electric and magnetic fields which he produced in a vacuum tube to measure the mass and charge of electrons as they flowed as beams of energy at much slower than light speed. His early model of the atom was called the "plum pudding model," because it depicted the atom as a "pudding" of positive charge embedded with particles of negative charge. The model served well to explain such phenomena as electrolysis and static electricity by presuming that the embedded negative charges could be detached from the positive pudding.

The New Zealand born, English physicist **Ernst Rutherford** (b. 1871; d. 1937) demonstrated that the **nucleus** of an atom was extremely compact. Illustration A shows a simplified version of the experiment performed by Rutherford to support his argument. Alpha radiation composed of positively charged helium nuclei was emitted from a shielded source and directed at a sheet of gold foil. Thomson predicted that the positively charged "pudding" centers of the gold atoms would deflect the similarly charged alpha particles. Instead, the vast majority of alpha particles passed easily through the foil producing visible flashes of light on a zinc sulfide screen. Rutherford concluded that the nuclei of the gold atoms had to be very small and compact in order for the alpha particles to pass through the foil without deflection. In 1913, Rutherford's gifted student, **Henry Moseley** (b. 1887; d. 1915), used X-ray spectroscopy to determine the number of positively charged protons occupying the nuclei of atoms. Moseley's new **atomic numbers** were used

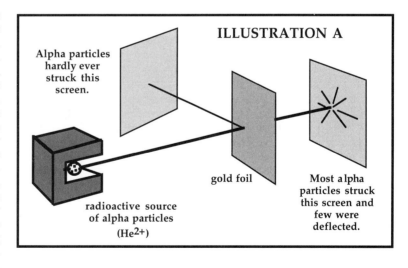

ILLUSTRATION A

Alpha particles hardly ever struck this screen.

gold foil

Most alpha particles struck this screen and few were deflected.

radioactive source of alpha particles (He^{2+})

to revise The Periodic Table of the Elements put forth by **Dmitri Mendeleev** (b. 1834; d. 1907). Moseley was killed in World War I at the Battle of Gallipoli.

Working with Rutherford in England, the Danish physicist **Niels Bohr** (b. 1885; d. 1962), explained the variety of light spectra emitted by the atoms of different elements using his **electron-shell model** of the atom. According to the Bohr model, electrons jumped from one electron-shell or "energy level" to another by absorbing or emitting electromagnetic energy. Electrons jumped to higher levels by absorbing discrete quantities of energy; and, fell to lower levels by emitting discrete quantities of energy. Many scientists contested Bohr's model of the atom, arguing that the negatively charged electrons should spiral into the positively charged nucleus as a result of the mutual attraction between the particles. Bohr's model presented a paradox. Bohr resolved the paradox by assuming that electrons could only occupy well-defined orbits and could not travel about the atom in continuous pathways.

CH7 Content Notes *(cont'd)*

Bohr declared that electrons made "quantum leaps" between energy levels *without traversing the intervening space* between the levels. His argument seemed outlandish to many at the time, but was supported by the work of the German physicist **Max Planck** (b. 1858; d. 1947). Planck's study of **black-body radiation** showed that electromagnetic energy was indeed emitted from atoms in discrete packets—or quanta—of energy and did not produce a continuous spectrum of light. These phenomena cannot be explained by the **classical mechanics** of **Sir Isaac Newton** (b. 1642; d. 1727) but can be understood when one views the atom according to the Bohr model. The work of Planck and Bohr served as the foundation for **quantum mechanics**: the new science of the atom founded by the German physicist **Werner Karl Heisenberg** (b. 1901; d. 1976). According to the **quantum-mechanical model**, or **cloud model**, of the atom, the movement of electrons is restricted to well-defined energy levels or "zones of probability." Whether an electron can be found in one energy or another—or at any given point within an energy level for that matter—is a matter of probability. The German born, American physicist, **Albert Einstein** (b. 1879; d. 1955), was revolted by this notion. He is quoted as having exclaimed to Bohr "God does not play dice with the universe!" To which Bohr is said to have responded: "Who are you to tell God what to do?" While much of quantum mechanics is counter intuitive, the theory is supported by sound laboratory research. In the last several decades, teams of physicists working at **particle accelerators** around the world have succeeded in smashing protons into smaller particles called **quarks**. A quark can be demonstrated to have a "fractional charge." And, in triplet combination they comprise the larger protons and neutrons of the atom. Despite the apparent complexity of the current model of the atom, the Bohr model is sufficient to explain the chemical interactions between atoms of different elements.

In Lesson #1, students will draw diagrams of the "plum pudding model," the "Bohr model," and the "cloud model" of the atom to show how models of atoms have evolved. They will be introduced to The Periodic Table of the Elements and shown how to calculate the number of neutrons in the nucleus of an atom using The Table.

In Lesson #2, students will construct models of Bohr atoms and learn to assign specific numbers of electrons to different energy levels.

In Lesson #3, students will diagram how atoms are transformed into ions by losing and gaining electrons.

In Lesson #4, students will diagram how atoms can form ionic and covalent bonds to make different chemical compounds.

ANSWERS TO THE HOMEWORK PROBLEMS

(1)

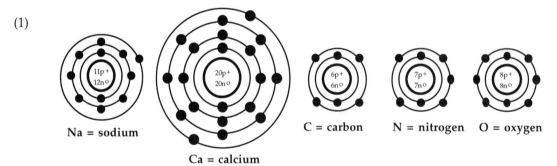

Na = sodium   Ca = calcium   C = carbon   N = nitrogen   O = oxygen

(2) Na (1 dot); C (4 dots), O (6 dots), P (5 dots), Mg (2 dots), Ca (2 dots), N (5 dots), Cl (7 dots), Si (4 dots), Kr (8 dots)

Be sure students write the chemical symbols correctly and put no more than 2 dots on any given side of that symbol.

ANSWERS TO THE END-OF-THE-WEEK REVIEW QUIZ

1. true	6. 1,840	11. A	16. F
2. electrons	7. true	12. B	17. H
3. neutrons	8. true	13. C	18. G
4. quarks	9. family	14. E	
5. electrons	10. true	15. D	

Li (1 dot)   Al (3 dots)   Ne (8 dots)

CH7 Fact Sheet

ATOMIC STRUCTURE

CLASSWORK AGENDA FOR THE WEEK

(1) Draw diagrams of atoms to show how models of atoms have evolved.
(2) Construct a model of a Bohr atom.
(3) Diagram how atoms can be transformed into ions.
(4) Diagram how atoms can form ionic or covalent bonds with other atoms.

What do the following have in common: a gorilla, an apple tree, your socks, the planet Mars, this piece of paper? Answer: They are all made of **atoms**. Even the ink in the dot atop the "i" in the word "ink" is made of atoms: about a trillion of them! That's 1,000,000,000,000 atoms (10^{12} atoms). It's hard to imagine that objects so large can be made of objects so small but it's a fact. The ancient Greeks guessed that atoms existed but no one knew for sure that they actually did until the beginning of the 20th century. The great American physicist **Albert Einstein** (b. 1879; d. 1955) convinced scientists that atoms must exist in 1905. He proved that the tiny movements of particles seen jumping around under a microscope—a phenomenon called **Brownian motion**—was the result of atoms hitting and jostling the particles to make them move.

The Greek philosopher **Democritus** (b. 460 B.C.; d. 370 B.C.) was the first to use the term "atoms" to describe these incredibly tiny units of matter. The English chemist **John Dalton** (b. 1766; d. 1844) developed the first **atomic theory of atoms** based on laboratory experiments. Today, scientists describe matter according to the **Atomic-Molecular Theory**. According to the Atomic-Molecular Theory of Matter, all matter is made of tiny particles that are in constant motion. The first clear estimate of the actual size of an atom was made by the American scientist **Irving Langmuir** (b. 1881; d. 1957) in 1917. He showed that an average atom was about one-billionth of a meter in diameter.

In the past century, scientists have discovered that an atom is made of three **subatomic particles**: protons, neutrons, and **electrons**. The English physicist **J. J. Thomson** (b. 1856; d. 1940) discovered the *negatively charged electron* in 1897. In 1911, the New Zealand-English physicist **Ernst Rutherford** (b. 1871; d. 1937) showed that atoms have a small but heavy center called a **nucleus**. The nucleus contains *protons* that have a *positive electrical charge* and a mass that is 1,840 times greater than that of an electron. Rutherford showed that the nucleus is extremely small compared to the size of the entire atom. If the nucleus of an atom were the size of a basketball, and you placed the ball on the 50-yard-line of a football field, then the tiny orbiting electrons would be as far away as the stands. Neutrons, which occupy the nucleus along with protons, have no electrical charge. Neutrons having about the same mass as protons were discovered in 1932 by the English physicist **James Chadwick** (b. 1891; d. 1974). Most recently, scientists have discovered that protons and neutrons are made of smaller particles of matter which they call **quarks**. Electrons belong to a group of subatomic particles called **leptons**. *The arrangement of electrons surrounding the nucleus gives an atom its chemical properties.*

In 1913, the Danish physicist **Niels Bohr** (b. 1885: d. 1962) showed that electrons orbit the nucleus in well-defined orbits like the planets of our solar system orbit the sun. In the **Bohr Model** of an atom, an **electron-shell diagram** shows the arrangement of *all of the electrons surrounding the nucleus* of an atom. Figure A shows a Bohr Model of a lithium atom. There are 3 protons and 4 neutrons in the nucleus and 3 electrons orbiting in two "electron-shells." Negatively charged electrons are attracted to

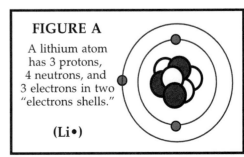

FIGURE A

A lithium atom has 3 protons, 4 neutrons, and 3 electrons in two "electrons shells."

(Li•)

the positively charged nucleus but are forced to orbit in different shells because their "like" electrical charges repel one another. The electrons in the "outermost electron-shell" are the only electrons involved in chemical reactions. Chemists use **electron-dot formulas** to show the number of electrons in the "outermost electron-shell." The "electron-dot formula" for lithium is shown in parentheses in Figure A. With some exceptions, *The Periodic Table of the Elements,* can tell you how many electrons occupy the outer shell of any atom. Each vertical column in *The Table* is called a **family** and the Roman numeral at the top of each family is called a **family number**. In families IA, IIA, IIIB, IVB, VB, VIB, and VIIB the family number tells how many electrons are in the atom's last shell. In family VIIIB, helium is the only atom with only two electrons in its outer shell. All of the other elements in family VIIIB have atoms with eight electrons in their outer shell. The Bohr model of atoms helps a scientist explain why chemical elements react the way they do during chemical reactions.

Homework Directions

1. Draw *electron-shell diagrams* for the atoms of the following elements: sodium, calcium, carbon, nitrogen, and oxygen.

2. Draw *electron-dot formulas* for the atoms of the following elements: sodium, carbon, oxygen, phosphorus, magnesium, calcium, nitrogen, chlorine, silicon, and krypton.

Assignment due: _____

_____   _____   ____/____/____
Student's Signature   Parent's Signature   Date

CH7 Lesson #1

ATOMIC STRUCTURE

Work Date: ____/____/____

LESSON OBJECTIVE

Students will draw diagrams of atoms to show how models of atoms have evolved.

Classroom Activities

On Your Mark!

Begin the lesson by tearing a piece of paper into smaller and smaller pieces. Ask: Assuming I had the most powerful microscope possible and the sharpest cutting tools, could I continue to divide this piece of paper forever? Or, would I eventually reach a particle that could not be divided any further: an indestructible particle? The class will be divided on this and for good reason. Scientists do not know the answer to this question. The **quark** is the smallest particle known and—according to the modern **Standard Model**—there are six different quarks each having its own "antiparticle." Remind students that **Democritus** (b. 460 B.C.; d. 370 B.C.) suggested the name **atom** to describe the indestructible particle that made up all of matter. Draw the diagrams in Illustration B to show students how the model of the atom has changed over the years. Have them copy your drawings into their notes on Journal Sheet #1. Point out that atoms were not seen until recently and that **Albert Einstein** (b. 1879; d. 1955) was the first to prove that atoms actually existed. He did this by explaining the phenomenon called **Brownian motion**. Brownian motion—first discovered by the Scottish botanist **Robert Brown** (b. 1773; d. 1858) in 1827—is the visible "jostling" of tiny particles in a transparent fluid. Students can observe the Brownian motion of very small iron dust particles under a microscope with a magnification of 400 or greater. If a powerful microscope is available have students refer to the *Microscope View of Brownian Motion* on Journal Sheet #1 and perform the following activity: (1) Mix some iron filings in water. (2) Place a drop of the solution onto a microscope slide. (3) Place the slide on the microscope stage and *carefully* lower the high power objective until the drop of water adheres to the objective lens. (4) Focus slowly up and down through the water droplet to see if you can observe some of the particles jostling around. (5) Any movement or jiggling of the scope will make the observation of Brownian motion difficult if not impossible. Even a running air conditioner will cause interfering vibration.

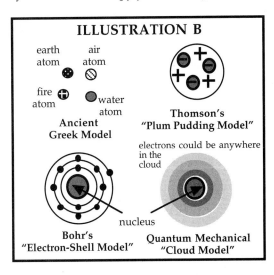

ILLUSTRATION B

earth atom · air atom ⊘

fire atom ⊕ · water atom ●

Ancient Greek Model

Thomson's "Plum Pudding Model"

electrons could be anywhere in the cloud

nucleus

Bohr's "Electron-Shell Model"

Quantum Mechanical "Cloud Model"

Get Set!

Copy and distribute *The Periodic Table of the Elements* provided in the Appendix and show students how to interpret the information in each "element box." Point out that the **atomic mass** of an element is the mass in grams of approximately 600 billion trillion atoms of that element. It is also the sum total of the number of **protons** and **neutrons** in the nucleus of an atom. The number of **electrons** in a **neutral atom** is equal to the number of protons. Show students how to calculate the number of neutrons in an atom by subtracting the atomic number from the atomic mass.

Go!

Have students practice interpreting the information on *The Periodic Table of the Elements* by completing Table A on Journal Sheet #1.

Materials

microscope with 400 magnification or greater, iron filings, microscope slides, water

CH7 JOURNAL SHEET #1

ATOMIC STRUCTURE

MICROSCOPE VIEW OF BROWNIAN MOTION

The large arrows show motion that is shared by the particles and is probably the result of fluid currents caused by jiggling the whole slide. The "random" zig-zag twirling motion of the particles, however, is Brownian motion: the result of continuous and random impacts by the "invisible" particles of the fluid.

TABLE A						
chemical element	chemical symbol	atomic mass	atomic number	no. of protons	no. of neutrons	no. of electrons

ATOMIC STRUCTURE

Work Date: ____/____/____

LESSON OBJECTIVE

Students will construct models of Bohr atoms.

Classroom Activities

On Your Mark!

Ask students to visualize the following model of an atom as you draw an electron shell diagram of a **sodium atom**: A sodium atom has 11 positively charged protons in its nucleus that will attract 11 electrons with negative electrical charges. The first 2 electrons attracted to the nucleus will occupy the first **electron-shell** or **energy level** as though they were racing around the surface of a ping-pong ball that has a marble inside it. Imagine the marble is the atom's nucleus and the surface of the ping-pong ball is the first electron-shell. The nucleus can still attract 9 more electrons. But as those electrons fall toward the nucleus they are repelled by the electrons already there. In order for the forces of attraction (to the nucleus) and repulsion (away from other electrons) to be balanced, the approaching electrons must reside in a higher electron-shell farther from the nucleus. Imagine the ping-pong-ball-marble atom placed inside a hollowed out orange. According to the complicated mathematical calculations made by **Niels Bohr** (b. 1885; d. 1962) and his colleagues, the second electron-shell can only hold up to 8 electrons before other approaching electrons are pushed to even higher energy levels. The last electron attracted to the nucleus of the sodium atom will occupy a third energy level as though our hollowed out orange-ping-pong-ball-marble atom were placed inside a basketball.

Get Set!

Draw Illustration C on the board and have students copy the drawing on Journal Sheet #2. Inform students that Bohr's model of the atom was the first to explain why atoms of different elements give off specific colors of light when heated to very high temperatures. Electrons leaping between energy levels emit specific frequencies of electromagnetic radiation. The formula for

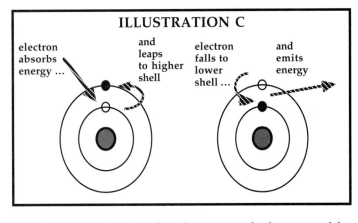

ILLUSTRATION C

electron absorbs energy … and leaps to higher shell / electron falls to lower shell … and emits energy

finding the amount of energy emitted by atoms is $E = h\nu$: where E=energy, ν=the frequency of the radiation, and h=**Planck's constant**. This formula was worked out by the German physicist **Max Planck** (b. 1859; d. 1947) who discovered that the amount of energy in all kinds of electromagnetic energy is always a multiple of Planck's constant. This means that all forms of electromagnetic energy is transmitted in tiny packets which Planck called "quanta."

Go!

Assist students in constructing the clay models of the atoms shown in Figure A on Journal Sheet #2. Advise them to space the electrons as far apart as they can within each electron-shell since they are repelling one another at all times. Explain that the maximum number of electrons that can fit in the electron-shells of atoms in the first 3 **periods** of *The Periodic Table of the Elements* is 2, 8, and 8, respectively. Show them how to draw Bohr **electron-shell diagrams** and write **electron-dot formulas** like those in Figure A.

Materials

three colors of clay, toothpicks or plastic straws cut into three sets of different lengths

CH7 JOURNAL SHEET #2

ATOMIC STRUCTURE

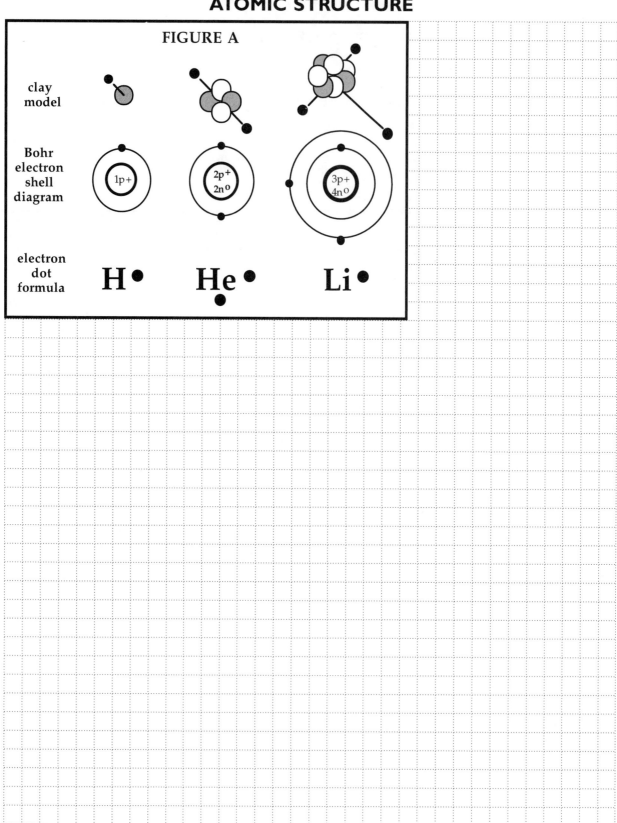

FIGURE A

clay model

Bohr electron shell diagram

electron dot formula

H •   He •   Li •

ATOMIC STRUCTURE

Work Date: ____/____/____

LESSON OBJECTIVE

Students will diagram how atoms can be transformed into ions.

Classroom Activities

On Your Mark!

After reviewing the structure of atoms according to the Bohr model, explain that electrons in the last electron-shell are "vulnerable." If atoms have less than 4 electrons in their outer shell they tend to "lose" them. Losing negatively charged electrons will leave an atom with more positively charged protons in the nucleus which cannot leave under ordinary circumstances. Changing the number of protons in the nucleus of an atom changes that atom to another element. This only occurs in **nuclear reactions** which are discussed in another unit of this volume. An atom with more protons than electrons becomes a **positive (+) ion**. Atoms with more than 4 electrons in their outer shell tend to "gain" electrons. This is because "gaps" remain in unfilled shells far from the nucleus allowing extra electrons to be attracted to the "exposed" protons of the nucleus. Outer-shells are most stable when they are filled. **Niels Bohr** (b. 1885; d. 1962) determined the number of electrons that would fill a given shell in a complex series of mathematical calculations that obey the **laws of electromagnetism** derived by the Scottish physicist **James Clerk Maxwell** (b. 1831; d. 1879). Atoms with more electrons than protons become **negative (-) ions**. Atoms with exactly 4 electrons in their outer shell are hesitant to lose or gain electrons. These atoms can share their electron-shells with other atoms.

Get Set!

Refer students to Figure B on Journal Sheet #3. Explain how hydrogen and beryllium lose electrons to become positive ions and how fluorine and oxygen gain electrons to become negative ions.

Go!

Assist students in writing the equations shown in Figure B until they understand how atoms can be transformed to ions.

Materials

Journal Sheet #3

CH7 JOURNAL SHEET #3

ATOMIC STRUCTURE

FIGURE B

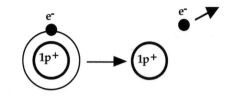

$$H\bullet \longrightarrow H^+ + e^-$$

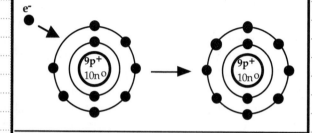

$$\bullet\ddot{\underset{\bullet\bullet}{F}}{:} + e^- \longrightarrow F^-$$

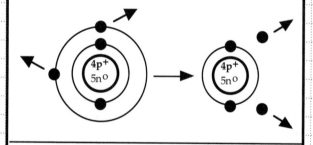

$$Be\,{:} \longrightarrow Be^{2+} + 2e^-$$

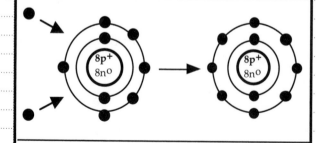

$$\ddot{\underset{\bullet\bullet}{:}O}{:} + 2e^- \longrightarrow O^{2-}$$

ATOMIC STRUCTURE

Work Date: ____/____/____

LESSON OBJECTIVE

Students will diagram how atoms can form ionic or covalent bonds with other atoms.

Classroom Activities

On Your Mark!

Review how atoms in families IA IIA, IIIB, VB, VIB and VIIB form ions. Remind students that atoms in family IVB form covalent bonds by sharing outer shells. Ask students to explain why atoms in family VIIIB are chemically "unreactive." Answer: The atoms of family VIIIB already have filled outer shells that make them stable.

Get Set!

Refer students to Figure C on Journal Sheet #4. Explain how a **positive hydrogen ion** is attracted to a **negative fluorine ion** to form hydrofluoric acid, an **ionic compound**. Explain how a carbon atom will attract four hydrogen atoms and share its outer shell to form methane gas, a **covalent compound**.

Go!

Assist students in drawing Bohr diagrams of ionic and covalent compounds (i.e., BeO, NaCl, CO_2, Si_2O). Remind students that atoms in Family IVB of The Periodic Table are the most likely to form covalent bonds.

ILLUSTRATION D

IONIC COMPOUND (beryllium oxide)

4p+   8p+

$$Be^{2+} + O^{2-} \longrightarrow BeO$$

COVALENT COMPOUND (carbon dioxide)

8p+   6p+   8p+

$$C + 2O \longrightarrow CO_2$$

*Note that the outer shells of the single carbon atom and each of the two oxygen atoms has 8 electrons free to roam their atom's outer shell; although carbon's electrons are limited to movement around the oxygen atoms. This is a simplified version of the true picture. In order to explain the complexity of interactions between atoms, Bohr's electron-shells are subdivided into oddly shaped "suborbitals" which give a more accurate description of how molecules form covalent bonds

---

Materials

Journal Sheet #4

CH7 JOURNAL SHEET #4

ATOMIC STRUCTURE

FIGURE C

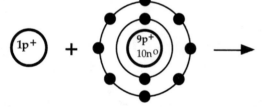

Ions are held together by attraction of opposite charges.

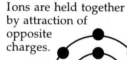

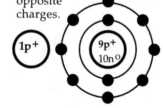

Hydrofluoric acid held together by ionic bonds:

$$H^+ + F^- \longrightarrow HF$$

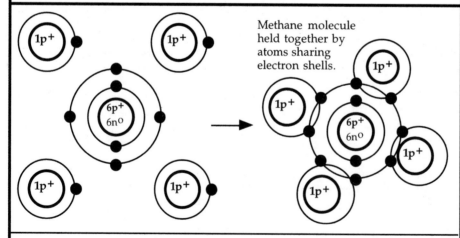

Methane molecule held together by atoms sharing electron shells.

Methane held together by covalent bonds:

$$C + 4H \longrightarrow CH_4$$

CH7 REVIEW QUIZ

Directions: Keep your eyes on your own work.
Read all directions and questions carefully.
THINK BEFORE YOU ANSWER!
Watch your spelling, be neat, and do the best you can.

CLASSWORK   (~40): _____
HOMEWORK   (~20): _____
CURRENT EVENT   (~10): _____
TEST   (~30): _____

TOTAL   (~100): _____
(A ≥ 90, B ≥ 80, C ≥ 70, D ≥ 60, F < 60)

LETTER GRADE: _____

TEACHER'S COMMENTS: _____
_____
_____

ATOMIC STRUCTURE

TRUE–FALSE FILL-IN: If the statement is true, write the word TRUE. If the statement is false, change the underlined word to make the statement true. *10 points*

_____ 1. <u>Protons</u> have a positive electrical charge.

_____ 2. <u>Neutrons</u> have a negative electrical charge.

_____ 3. <u>Electrons</u> have a neutral electrical charge.

_____ 4. Protons and neutrons are made of <u>chalk</u>.

_____ 5. Outer-shell <u>protons</u> give an atom its chemical properties.

_____ 6. Protons are about <u>100</u> times more massive than electrons.

_____ 7. An "electron-<u>shell diagram</u>" shows all of the electrons surrounding the nucleus of an atom.

_____ 8. An "electron-<u>dot formula</u>" shows only the electrons farthest from the nucleus.

_____ 9. The <u>atomic</u> number tells how many electrons are in an atom's last shell.

_____10. Electrons belong to a group of subatomic particles called <u>leptons</u>.

MATCHING: Match the letter of the scientific accomplishment on the right to the name of the scientist who achieved it. *8 points*

_____ 11. Einstein   (A) proved that atoms exist by explaining Brownian motion

_____ 12. Democritus   (B) gave the atom its name

_____ 13. Dalton   (C) developed the first atomic theory based on lab research

_____ 14. Langmuir   (D) discovered the electron

_____ 15. Thomson   (E) measured the size of atoms

_____ 16. Chadwick   (F) discovered the neutron

_____ 17. Rutherford   (G) showed that electrons have well-defined orbits

_____ 18. Bohr   (H) proved the nucleus was small compared to the whole atom

PROBLEMS

Directions: Using a *Periodic Table of the Elements,* draw an Electron-Shell Diagram <u>and</u> an Electron-Dot Formula for the elements lithium (Li), aluminum (Al), and neon (Ne). *12 points*

LITHIUM

| electron-shell diagram | electron-dot formula |

ALUMINUM

| electron-shell diagram | electron-dot formula |

NEON

| electron-shell diagram | electron-dot formula |

THE PERIODIC TABLE
OF THE ELEMENTS

TEACHER'S CLASSWORK AGENDA AND CONTENT NOTES

Classwork Agenda for the Week

1. Students will be able to explain how elements are arranged on The Periodic Table.

2. Students will be able to compare and contrast the physical and chemical properties of families on The Periodic Table.

3. Students will prepare a group report on the elements of a chemical family.

4. Students will present a group report on the elements of a chemical family.

Content Notes for Lecture and Discussion

In 1789, the French chemist **Antoine Laurent Lavoisier** (b. 1743; d. 1794) published *Traité Élémentaire de chimie* (*The Thirty Chemical Elements*) in which he listed and described the chemical and physical properties of the thirty known elements of his time. Other chemists were similarly anxious to find an organized way of classifying the elements after Lavoisier defined an element as a "substance that could not be further analyzed." In 1753, the Swedish naturalist **Carolus Linnaeus** (b. 1707; d. 1778) simplified the classification of living species by introducing his nomenclature for organizing the living kingdoms, and chemists sought the practical and theoretical benefits of having an organized system of classification for the natural elements. With the development of newer methods of chemical analysis resulting from studies in electrochemistry and spectroscopy more elements were on the verge of discovery.

The **atomic theory** of **John Dalton** (b. 1766; d. 1844) served as the impetus to find an appropriate classification system, the theory suggesting that it might be possible to compare atoms according to their relative weights. But it was not until 1807 that the English chemist **Humphry Davy** (b. 1778; d. 1829) showed that the elements had an electrical "affinity" for one another that determined their reactivity. Using electrolysis, Davy discovered the elements barium, boron, magnesium, potassium, sodium, and strontium within a period of two short years. His experiments allowed him to list the known and newly discovered elements in an "**electrochemical series**" which was—in effect—a scale of relative chemical affinities. In the 1860s, several chemists stumbled upon the idea that the chemical and physical properties of the elements were "periodic" and that it might be possible to group the elements together in **families of elements** whose "siblings" shared similar characteristics. But the English chemist **John Newlands** (b. 1837; d. 1898) was the first to propose the idea of a "periodic law" publicly. Despite ridicule by the members of the Chemical Society of London, Newlands' **Law of Octaves** was given firm and lasting confirmation in 1869 by the work of **Dmitri Mendeleev** (b. 1834; d. 1907) and the German chemist **Lothar Meyer** (b. 1830; d. 1895). In 1870, Meyer published a graph which clearly showed a periodic relationship between the relative atomic volumes and atomic weights of the known elements; but his work was overshadowed by the uncanny intuition of Mendeleev. The power of Mendeleev's arguments derived from his predictions of the precise characteristics of the—as yet—undiscovered elements: gallium, scandium, and germanium. Over the next several decades, versions of Mendeleev's "periodic table" were revised; and, the elements—listed in order of increasing atomic mass from the upper left of the chart toward the lower right—were arranged in **vertical families** and **horizontal periods**.

CH8 Content Notes *(cont'd)*

The elucidation of the structure of the atom by the physicist **Ernst Rutherford** (b. 1871; d. 1937) allowed **Henry Moseley** (b. 1887; d. 1915)—working in Rutherford's laboratory—to show that **atomic number** and not **atomic mass** was the determinant factor in making sense of the periodic law. When atoms were placed in order of their atomic number (i.e., a measure of the number of positively charged protons in the nucleus of the atom), all of the elements "fell into line" on The Periodic Table in agreement with empirical laboratory data. The **electron-shell model** of the atom, proposed by the Danish physicist **Niels Bohr** (b. 1885; d. 1962), allowed chemists to explain the absolute reactivities of the elements.

In Lesson #1, students will show how hypothetical substances can be arranged in a periodic way analogous to the arrangement of the elements of The Periodic Table.

In Lesson #2 and Lesson #3 students will use library resources to prepare a rough and final draft of the elements in a chemical family.

In Lesson #4 students will present a group report on the elements of a chemical family and take notes on the presentations of other groups.

ANSWERS TO THE HOMEWORK PROBLEMS

Answers will vary, but students should demonstrate an understanding of the definitions obtained from the dictionary in making a clear distinction between the elements of the families listed.

ANSWERS TO THE END-OF-THE-WEEK REVIEW QUIZ

1. C	6. C	#11 through #15: Answers will vary in order and content
2. A	7. E	
3. A	8. D	
4. B	9. A	
5. A	10. C	

CH8 FACT SHEET

THE PERIODIC TABLE OF THE ELEMENTS

CLASSWORK AGENDA FOR THE WEEK

(1) Explain how elements are arranged on The Periodic Table.
(2) Compare and contrast the physical and chemical properties of families on The Periodic Table.
(3) Prepare a group report on the elements of a chemical family.
(4) Present a group report on the elements of a chemical family.

In 1864, an English industrial chemist named **John Alexander Newlands** (b. 1837; d. 1898) made a list of the chemical elements known to him. He listed the elements by their mass from the lightest to the heaviest. Upon studying his list he noticed something interesting. Every eighth element—starting from any element on the list—had many of the same chemical and physical characteristics as the first in the series of eight. This "periodicity" reminded him of the way musical notes sound on a piano. The middle "c" on a piano scale sounds much the same as the eighth note one octave away from it. Newlands noted that the element sodium (Na) reacts violently when mixed with water as does potassium (K). On Newland's list, potassium was listed eight elements away from sodium. The element magnesium (Mg) has many of the same properties as calcium (Ca) which is eight elements away from magnesium. Two years later Newlands proposed his **Law of Octaves** to the Chemical Society of London and was told that his idea was silly. Three years after Newland's disastrous presentation at the Chemical Society, a Russian chemist named **Dmitri Mendeleev** (b. 1834; d. 1907) showed that Newlands' idea was exactly correct. The chemical elements could be grouped into **families of chemical elements** having similar physical and chemical properties. Mendeleev also showed that the order of the arrangement of the chemical elements followed a definite pattern just as Newlands had discovered.

Mendeleev published his own chart in 1869. He arranged the elements just as Newlands had done according to their increasing atomic mass. Then, he put them in rows according to the way they "reacted" when mixed with other elements. Today, the modern **Periodic Table of the Elements** is arranged in horizontal rows called **periods** and vertical columns called **families**. From the "gaps" in his chart Mendeleev predicted the properties of gallium, scandium, and germanium which had not yet been discovered. All three elements were found within the next 20 years and had exactly the properties Mendeleev predicted they would. Since Mendeleev's time, other elements have been added to the chart and the **periodic law of chemistry** remains a valuable tool for understanding how elements behave. The periodic law states that all of the elements are related according to their atomic masses in a single orderly system. Since Mendeleev's *Periodic Table of the Elements* was first published, many more elements have been discovered and added to include more than one hundred chemical elements.

Elements in the same family have very similar chemical properties. For example, the element sodium (Na) is a highly reactive "light metal." It produces a violent chemical reaction when mixed with water. Lithium (Li) is in the same family as sodium. Lithium also reacts violently when mixed with water. Sodium and lithium are both **alkali metals** in Family IA (or 1). Chlorine (Cl), bromine (Br), and iodine (I) are in a different family called the **halogens**: Family VIIB (or 17). All three substances are poisonous and react violently when mixed with any of the alkali metals. The **noble gases** in Family VIIIB (or 18) are unreactive gases. It is extremely difficult to get any of the elements in this family to react with other chemical substances.

CH8 Fact Sheet *(cont'd)*

In the unit entitled Atomic Structure, you learned that atoms have a positively charged **nucleus** and negatively charged **electrons** that orbit it. Today we know that the manner in which an atom reacts with other atoms depends on the number of electrons in the "outermost" electron-shell of that atom. If you draw electron-shell diagrams of atoms in *The Periodic Table* you will notice that elements in the same chemical families have the same number of electrons in their outer shell. These outer shell electrons are called **valence electrons**.

Homework Directions

1. Copy the dictionary definition of the following terms: metal, brittle, ductile, boiling, and inert.
2. Refer to *The Periodic Table* given to you by your instructor and to the box titled "Physical Properties." Write a paragraph that contrasts the properties of the elements in the listed families **without using** the following terms: metal, brittle, ductile, boiling, or inert.

Assignment due: _____

_____   _____   ____/____/____
Student's Signature   Parent's Signature   Date

THE PERIODIC TABLE OF THE ELEMENTS

Work Date: ____/____/____

LESSON OBJECTIVE

Students will be able to explain how elements are arranged on The Periodic Table.

Classroom Activities

On Your Mark!

Discuss the concept of "periodicity." Examples of "periodic" events include the change of the seasons, the need to eat and sleep at regular intervals, and the reproductive cycles of animals (i.e. bird migration and nesting behavior). Agree on the following definition: "Periodicity is the regular cyclical appearance of a phenomenon." Give a brief lecture of the history of *The Periodic Table of the Elements* using the Teacher's Agenda and Content Notes. Students may take notes on the most important scientists involved in the development of the chart in the margins of Journal Sheet #1. Stress the point that the chart is used to give chemists a simplified way of identifying the elements. Define an element as **Antoine Laurent Lavoisier** (b. 1743; d. 1794) defined it: "An **element** is a substance that cannot be further analysed (i.e., it is in its simplest form). Review the concepts of **atomic mass** and **atomic number**. Have students refer to *The Periodic Table of the Elements* and tell you the atomic masses and numbers of several elements. Point out that elements are listed from left to right across the table in order of increasing atomic number.

Get Set!

Examine the *problem* on Journal Sheet #1 making sure that students understand the classification criteria used by "the stranded space explorer."

Go!

Allow students time to fill in the chart. Discuss the most logical arrangement shown in Illustration A. Point out that the "tiny circles" attached to each "larger circle" may represent an additional chemical

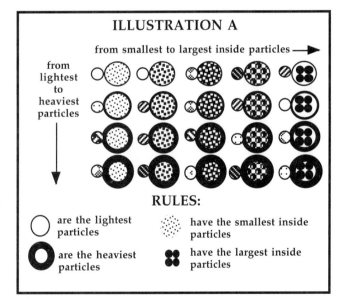

ILLUSTRATION A

from smallest to largest inside particles ⟶

from lightest to heaviest particles ↓

RULES:

◯ are the lightest particles

◉ are the heaviest particles

⬤ have the smallest inside particles

⬤ have the largest inside particles

or physical property of an element; but students need only pay attention to the large circles (i.e., characteristics in common) in completing the table.

Materials

Journal Sheet #1

CH8 JOURNAL SHEET #1

THE PERIODIC TABLE OF THE ELEMENTS

PROBLEM

On an alien planet in the Mysterio Galaxy a stranded space explorer discovers a rock of unusual composition. Aware that his understanding of this new environment can mean the difference between life and death, the explorer analyzes the rock. He discovers that it contains twenty (20) unique elements. His experiments also show the following:

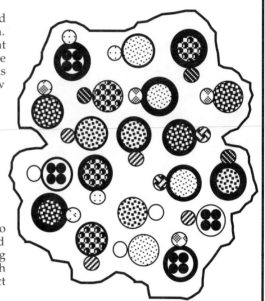

○   are the lightest particles

⬤   are the heaviest particles

∴   have the smallest inside particles

❀   have the largest inside particles

In order to understand how these elements are related to one another, he groups them according to their mass and the sizes of the particles that comprise their centers. Using the results of the space explorer's experiments, draw each of the 20 elements shown in the rock sample in its correct box on THE TABLE OF MYSTERIO ELEMENTS.

THE TABLE OF MYSTERIO ELEMENTS

from smallest to largest inside particles ⟶

from lightest to heaviest particles ↓				

THE PERIODIC TABLE OF THE ELEMENTS

Work Date: _____/_____/_____

LESSON OBJECTIVE

Students will be able to compare and contrast the physical and chemical properties of families on The Periodic Table.

Classroom Activities

On Your Mark!

Arrange with the librarian to allow your class to visit the library for purposes of research. In class, refer to *The Periodic Table of the Elements* and discuss the "Physical Properties" of the elements listed in the similarly titled box.

Get Set!

Remind students that while the elements of a family have chemical and physical properties in common each element is unique, having some individual properties as well. Refer to Journal Sheet #2 and give students the following electron-shell arrangements for the elements so that students can complete the Bohr Model of their atom. Divide the class into groups so that each group is responsible for gathering information on several chemical elements within a particular family.

PERIOD	Maximum number of electrons at each energy level
1	2
2	2 - 8
3	2 - 8 - 8
4	2 - 8 - 18 - 8
5	2 - 8 - 18 - 18 - 8
6	2 - 8 - 18 - 32 - 18 - 8
7	2 - 8 - 18 - 32 - 32 - 11 - 2

Go!

Lead students to the library to begin gathering information on the chemical and physical properties of their chosen element. Instruct them to use Journal Sheet #2 as a rough draft in preparing their report.

Materials

Journal Sheet #1, school library facility or Internet access

CH8 JOURNAL SHEET #2

THE PERIODIC TABLE OF THE ELEMENTS

BOHR MODEL OF

_____

CHEMICAL SYMBOL

_____

FAMILY NAME AND NUMBER

Directions: Use JOURNAL SHEET #2 at the library to prepare a rough draft of a brief report on your chosen element. Include facts about the chemical and physical properties of the element, when (and by whom) it was discovered, and where it can be found. On JOURNAL SHEET #3 you will use this information to prepare a neat final draft that you will share with your classmates.

THE PERIODIC TABLE OF THE ELEMENTS

Work Date: ____/____/____

LESSON OBJECTIVE

Students will prepare a group report on the elements of a chemical family.

Classroom Activities

On Your Mark!

Answer any questions students may have about their individual "elements reports."

Get Set!

Escort students to the library.

Go!

Instruct students to use Journal Sheet #3 as a final draft in preparing their report. Their final draft should be neat enough to display for the entire class to examine.

Materials

Journal Sheet #3, school library facility or Internet access

CH8 JOURNAL SHEET #3

THE PERIODIC TABLE OF THE ELEMENTS

BOHR MODEL OF

_____

**CHEMICAL
SYMBOL**

_____

**FAMILY NAME
AND NUMBER**

Directions: Use JOURNAL SHEET #3 to prepare a neat final draft of your "element report" using the information you gathered on JOURNAL SHEET #2. Share this information with your classmates in the manner described by your instructor.

CH8 Lesson #4

THE PERIODIC TABLE OF THE ELEMENTS

Work Date: ____/____/____

LESSON OBJECTIVE

Students will present a group report on the elements of a chemical family.

Classroom Activities

On Your Mark!

Display the final reports of each student in the class. Arrange the reports as they would appear on *The Periodic Table of the Elements* (i.e., in vertical families and horizontal periods).

Get Set!

Give students an opportunity to examine the display of reports and fill in the information gathered about the elements of other chemical family groups.

Go!

Give them time to report back to their own groups to compare their findings.

Materials

Journal Sheet #4

CH8 JOURNAL SHEET #4

THE PERIODIC TABLE OF THE ELEMENTS

FAMILY NAME AND NUMBER	FAMILY NAME AND NUMBER
_____	_____
FAMILY NAME AND NUMBER	**FAMILY NAME AND NUMBER**
_____	_____
FAMILY NAME AND NUMBER	**FAMILY NAME AND NUMBER**
_____	_____
FAMILY NAME AND NUMBER	**FAMILY NAME AND NUMBER**
_____	_____

CH8 Review Quiz

Directions: Keep your eyes on your own work.
Read all directions and questions carefully.
THINK BEFORE YOU ANSWER!
Watch your spelling, be neat, and do the best you can.

CLASSWORK	(~40): _____
HOMEWORK	(~20): _____
CURRENT EVENT	(~10): _____
TEST	(~30): _____
TOTAL	(~100): _____

(A ≥ 90, B ≥ 80, C ≥ 70, D ≥ 60, F < 60)

LETTER GRADE: _____

TEACHER'S COMMENTS: _____

_____

THE PERIODIC TABLE OF THE ELEMENTS

MULTIPLE CHOICE: Choose the letter of the word or phrase that best completes the sentence or answers the question. *20 points*

_____ 1. Who discovered the Law of Octaves?
   (A) Albert Einstein   (D) Sir Isaac Newton
   (B) Niels Bohr   (E) Dmitri Mendeleev
   (C) John Newlands

_____ 2. Horizontal rows on *The Periodic Table of the Elements* are called ___?___.
   (A) periods   (D) classes
   (B) families   (E) compounds
   (C) elements

_____ 3. Vertical columns on *The Periodic Table of the Elements* are called ___?___.
   (A) periods   (D) classes
   (B) families   (E) compounds
   (C) elements

_____ 4. How many elements can scientists classify on *The Periodic Table of the Elements* today?
   (A) about 10   (D) about 100,000
   (B) about 100   (E) about 1,000,000
   (C) about 1,000

_____ 5. Which elements produce the most violent chemical reactions when mixed with water?
   (A) alkali metals   (D) transition elements
   (B) alkaline metals   (E) metalloids
   (C) halogens

_____ 6. Which family includes highly reactive, poisonous gases?
   (A) alkali metals   (D) noble gases
   (B) alkaline metals   (E) metalloids
   (C) halogens

_____ 7. Which family includes unreactive gases?
   (A) alkali metals   (D) nonmetals
   (B) alkaline metals   (E) noble gases
   (C) halogens

_____ 8. Which of the following can be found in the nucleus of an atom?
     (A) protons   (D) both A and B
     (B) neutrons   (E) both B and C
     (C) electrons

_____ 9. Which electrons are always lost, gained, or shared in chemical reactions?
     (A) valence electrons   (D) electrons in the nucleus
     (B) second shell electrons   (E) all of the above
     (C) first shell electrons

_____ 10. Which is the best way to find out the number of neutrons in an atom when looking at *The Periodic Table of the Elements?*
     (A) find the atomic number   (D) add protons and electrons
     (B) find the atomic mass   (E) subtract atomic mass from
     (C) subtract atomic number   atomic number
          from atomic mass

List the names of five chemical families that you learned about this week in class. *10 points*

11. the _____ family

12. the _____ family

13. the _____ family

14. the _____ family

15. the _____ family

_____   _____   ____/____/____
   Student's Signature   Parent's Signature   Date

TYPES OF CHEMICAL REACTIONS

TEACHER'S CLASSWORK AGENDA AND CONTENT NOTES

Classwork Agenda for the Week

1. Students will distinguish between a synthesis and a decomposition reaction.
2. Students will perform a decomposition reaction that liberates explosive oxygen gas.
3. Students will capture an explosive gas as the product of a single displacement reaction.
4. Students will form a precipitate as the product of a double displacement reaction.

Content Notes for Lecture and Discussion

Any chemical change or **chemical reaction** can be described by a **chemical equation**. A chemical equation consists of **chemical symbols** and **chemical formulas** that detail the composition of a reaction's **reactants** and **products**. The chemical formulas ascribed to the vast variety of known substances are derived from the empirical studies conducted by thousands of chemists over the past several hundred years. In early experiments, for example, **Antoine Laurent Lavoisier** (b. 1743; d. 1794) found that a given weight of mercury combined with a given weight of oxygen formed a specific amount of mercuric oxide. Tin and lead combined with oxygen in a similarily precise fashion. A chemical formula, therefore, came to represent the ratio of elements within a compound: *a ratio that is constant for any given substance*. This principle constitutes the **law of definite proportions** suggested by Lavoisier's work and given final validation by the work of Belgian chemist **Jean Servais Stas** (b. 1813; d. 1891) and American chemist **Theodore William Richards** (b. 1868; d. 1928). Stas was among the first to make accurate determinations of the relative atomic weights of elements and Richards refined those measurements to account for the existence of isotopes. **Isotopes** are atoms with the same number of protons (i.e., equal positive charge) but different numbers of neutrons to account for the atoms' dissimilarity in mass. A summary of the "law of definite proportions" can be stated as follows: *A sample of any compound, regardless of its source, contains the same elements combined in a definite proportion by weight*. The lessons of this unit provide an excellent opportunity to review this law during lecture as well as the correct use of chemical symbols, formulas and equations.

In reviewing the concept of a **coefficient**—as you write balanced chemical equations to introduce students to the different types of chemical reactions—point out that the number of molecules involved in any chemical reaction is enormous. The Italian physicist **Amadeo Conte di Quaregna Avogadro** (b. 1776; d. 1856), called the "father of physical chemistry," first defined the concept of the **mole**. He determined that equal volumes of gases—at equal temperature and pressure—have the same number of molecules. One mole of any substance is Avogadro's number of molecules or 6.02×10^{23} molecules (i.e., 602,000,000,000,000,000,000,000 molecules). The word "molecule" means "a miniscule part of a mole." Careful experiments performed by chemists of the 18th and 19th centuries have since verified Avogadro's work as well as the **Law of Conservation of Mass**. As a result of this voluminous work, it is clear that in any chemical reaction there is no loss of matter. The atoms involved in a chemical reaction are rearranged but never lost. **Albert Einstein** (b. 1879; d. 1955) joined the **Law of Conservation of Mass** to the **Law of Conservation of Energy** with his famous formula **$E=mc^2$** by asserting that energy and matter are equivalent. In presenting the general and specific formulas and equations used by chemists to describe the types of chemical reactions, emphasize that matter and energy are neither created nor destroyed but merely rearranged to give us the fantastic variety of chemical substances we use in our every day lives.

CH9 Content Notes *(cont'd)*

In Lesson #1, students will oxidize a metal to show how a **synthesis reaction** occurs. Emphasize that oxygen can combine easily with many different substances. A demonstration of a **decomposition reaction** is also provided if materials are available.

In Lesson #2, students will perform a **decomposition reaction** and discover that two liquids that "look the same" (i.e., water and hydrogen peroxide) can have demonstrably different chemical properties.

In Lesson #3, students will capture an explosive gas as the product of a **single displacement reaction** and experience—first hand—the volatility of an explosive gas.

In Lesson #4, students will form a **precipitate** as the product of a **double displacement reaction**.

ANSWERS TO THE HOMEWORK PROBLEMS

(A) synthesis; (B) decomposition; (C) single displacement; (D) double displacement; (E) decomposition; (F) single displacement; (G) single displacement; (H) double displacement; (I) decomposition; (J) single displacement. Explanations will vary but should show the student's grasp of the fact that atoms are rearranged in chemical reactions.

ANSWERS TO THE END-OF-THE-WEEK REVIEW QUIZ

1. true
2. reactants
3. products
4. true
5. coefficient

6. true
7. exothermic
8. decomposition
9. decomposition
10. true

11. true
12. true
13. double displacement
14. true
15. compound

16. B
17. D
18. A
19. C

CHEMICAL EQUATION: $2H_2O \rightarrow 2H_2 + O_2$

CH9 FACT SHEET

TYPES OF CHEMICAL REACTIONS

CLASSWORK AGENDA FOR THE WEEK

(1) Distinguish between a synthesis and a decomposition reaction.
(2) Perform a decomposition reaction that liberates explosive oxygen gas.
(3) Capture an explosive gas as the product of a single displacement reaction.
(4) Form a precipitate as the product of a double displacement reaction.

Chemists use symbolic sentences called **chemical equations** to describe how elements and compounds behave when mixed. The substances mixed together in a chemical reaction are called the **reactants**. The substances produced in a chemical reaction are called the **products**. There are four basic classes of chemical reactions: *synthesis* reactions, *decomposition* reactions, *single displacement* reactions, and *double displacement* reactions.

In a **synthesis** reaction individual substances combine to form a single new substance. Imagine that "X" and "Y" are pure chemical elements. A chemical reaction describing a synthesis reaction involving elements "X" and "Y" might look like this:

$$2X_2 + Y_2 \rightarrow 2X_2Y$$

Remember that the number **2** in front of the X_2 and the X_2Y is called a "coefficient." A coefficient tells the chemist how many molecules of a substance are involved in any chemical reaction. In the chemical equation shown above, 2 molecules of X_2 react with 1 molecule of Y_2 to form 2 molecules of X_2Y. The small $_2$ written after each **X** or **Y** in the equation to the left or right of the arrow means there are 2 atoms of that element in each molecule containing that substance. The synthesis of water from oxygen and hydrogen gas is described by the following chemical equation: $2H_2 + O \rightarrow 2H_2O$. The synthesis of water is an **exothermic reaction** that releases an enormous amount of energy. This particular chemical reaction was responsible for the explosive destruction of the giant airship *Hindenburg* in 1937 and the space shuttle *Challenger* in 1986.

In a **decomposition** reaction, a compound is split into smaller chemical units. The general equation for a decomposition reaction might look like the following:

$$2X_2Y \rightarrow 2X_2 + Y_2$$

The electrolysis of water to form oxygen and hydrogen gas is a decomposition reaction. Electrolysis was the method used by the French chemists **Pierre Laplace** (b. 1749; d. 1827) and **Antoine Laurent Lavoisier** (b. 1743; d. 1794) to show that water was a compound and not an element as the ancient Greeks had once thought. The electrolysis of water is an **endothermic reaction** because electrical energy is absorbed by water molecules to produce individual oxygen and hydrogen molecules.

In a **single displacement** reaction, the atoms of an element "displace" atoms of another element that is already part of a compound. The general chemical equation for a single displacement reaction might look like the following:

$$X + YZ \rightarrow Y + XZ$$

In the example above, **X** displaces **Y**—which is already bonded to **Z** in compound **YZ**—to form the new compound **XZ**. This type of chemical reaction can be used to produce hydrogen gas by mixing an acid with a metal.

CH9 Fact Sheet *(cont'd)*

In a **double displacement** reaction, atoms in different compounds "displace" one another to form new compounds. The general chemical equation for a double displacement reaction might look like the following:

$$AB + CD \rightarrow AD + BC$$

In the example above, atom **A**—in compound **AB**—displaces atom **C**—in compound **CD**—to form the new compounds **AD** and **BC**. The most commonly used reaction of this class of chemical reactions is the **acid-base reaction** in which an acid is used to neutralize a base. This type of chemical reaction takes place when you take an antacid to cure an upset stomach.

Homework Directions

Directions: Which class of chemical reactions best describes each of the following chemical equations? Explain your answer in one or two sentences.

(A) $2H_2 + O_2 \rightarrow 2H_2O$

(B) $2KClO_3 \rightarrow 2KCl + 3O_2$

(C) $2Na + 2H_2O \rightarrow 2H_2 + 2NaOH$

(D) $NaOH + HCl \rightarrow NaCl + H_2O$

(E) $2HgO \rightarrow 2Hg + O_2$

(F) $2Ca + 2H_2O \rightarrow 2H_2 + 2CaO$

(G) $Zn + 2HCl \rightarrow 2H_2 + ZnCl_2$

(H) $Al_2(SO_4)_3 + 3Ca(OH)_2 \rightarrow 2Al(OH)_3 + 3CaSO_4$

(I) $CaCO_3 \rightarrow CaO + O_2$

(J) $Fe + H_2SO_4 \rightarrow H_2 + FeSO_4$

Assignment due: _____

116

TYPES OF CHEMICAL REACTIONS

Work Date: ____/____/____

LESSON OBJECTIVE

Students will distinguish between a synthesis and a decomposition reaction.

Classroom Activities

On Your Mark!

Begin discussion by asking students to define the term "synthesis." Point out that much of the clothing we wear today is "synthetic material." What does this mean? Mention that the term is commonly used to refer to materials that are "manmade." In chemistry a **synthesis reaction** is a reaction in which atoms and molecules are joined to make larger molecules. Explain that oxygen gas can easily combine with a variety of substances in a special type of synthesis reaction called an <u>oxidation</u> reaction. Refer students to Figure A on Journal Sheet #1 and explain the chemical equation shown below:

$$4Fe + 3O_2 \rightarrow 2\ Fe_2O_3$$
iron   oxygen   rust

Point out that this equation summarizes the synthesis reaction that will take place over the next few days in the experiment shown in Figure A. On the board, write the general equation for a synthesis reaction shown below and have students copy the equation on Journal Sheet #1. Review the meaning of the symbols in the equation. Compare that equation to the general equation for a decomposition reaction. Ask students to explain the difference between a synthesis reaction and a decomposition reaction merely by interpreting the two equations. Perform the demonstration shown in Illustration A and point out that synthesis is the opposite of decomposition.

$$\text{synthesis } X + Y \rightarrow XY \quad \textit{versus} \quad XY \rightarrow X + Y \text{ decomposition}$$

ILLUSTRATION A

<u>Directions</u>: (1) Place a tablespoon of copper (II) carbonate ($CuCO_3$) in a ceramic crucible. (2) Secure the crucible in a ring above a Bunsen burner on a ring stand. (3) Heat the contents of the crucible and note the color change of the contents. The decomposition reaction that has taken place can be summarized as follows:

$$CuCO_3 \longrightarrow CuO + CO_2$$

SAFETY PRECAUTIONS
Wear goggles to protect your skin and eyes against splattering. Do not touch any part of the equipment without heat-resistant gloves or tongs.

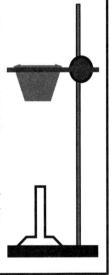

Get Set!

Refer students to their Fact Sheet and review the meanings of the symbols in the general synthesis and decomposition reactions shown there. Write the chemical equations for the synthesis and decomposition of water and contrast the two equations.

$$2H_2 + O_2 \rightarrow 2H_2O \qquad \textit{versus} \qquad 2H_2O \rightarrow 2H_2 + O_2$$
synthesis   decomposition

Go!

Have students set up the oxidation of iron demonstration shown in Figure A and perform the electrolysis of water experiment just as they did in Lesson #3 of CH6—*Elements, Molecules, and Compounds*. This time, have them write a paragraph that answers the question posed in Figure B.

Materials

ring stands and clamps, test tubes, 100 ml or 250 ml beakers, insulated copper wire, water, dilute sulfuric acid, D-cell batteries, switches, alligator clips, tape or battery holders, crucible, steel wool, detergent, microscope, copper (II) carbonate, Bunsen burner, goggles, gloves, tongs

CH9 JOURNAL SHEET #1

TYPES OF CHEMICAL REACTIONS

FIGURE B

Directions: (1) Fill a 100 ml beaker with water. (2) Add a medicine dropper full of the acid solution given to you by your instructor. AVOID SPLATTERING THE ACID. (3) Fill two small test tubes with water and without allowing air to enter the tubes invert them into the beaker. (4) Construct the set-up shown. (5) Insert the bare electrodes into the inverted test tubes. (6) Close the switch and allow the hydrogen gas to accumulate at the negative electrode and the oxygen gas to accumulate at the positive electrode. (7) Open the switch to shut off the current. (8) Carefully lift the hydrogen-filled test tube out of the water. Keeping it in an inverted position quickly cover the opening with your thumb. (9) Have your assistant light a match. (10) Remove your thumb from the test tube at the same instant your assistant puts the match flame to the opening. Carefully observe the inside of the test tube after the hydrogen "pops"! (11) Write a paragraph that answers the following questions: Which part of this demonstration involved a synthesis reaction? A decomposition reaction?

positive (+) electrode   negative (-) electrode

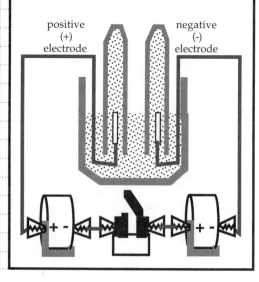

FIGURE A

Directions: (1) Examine small flakes of steel wool under a microscope and record their color. (2) Pour some water into a beaker and place it on a ring stand. (3) Wash and rinse a ball of steel wool with detergent and press it down to the bottom of a test tube. (4) Invert the test tube into the beaker and secure it as shown. Make sure the rim of the test tube is not touching the bottom of the beaker. (5) Examine the steel wool without removing it from the test tube every day for the next several days and record your comments. (6) After several days of observation, remove the steel wool and examine flakes of the steel wool under a microscope. (7) Compare their present color to the color they had at the start of this demonstration.

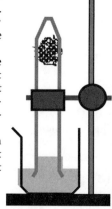

TYPES OF CHEMICAL REACTIONS

Work Date: ____/____/____

LESSON OBJECTIVE

Students will perform a decomposition reaction that liberates explosive oxygen gas.

Classroom Activities

On Your Mark!

Review the characteristics of a **decomposition reaction**. Explain that in most cases, energy is absorbed by a compound to make it decompose. Decomposition reactions are generally **endothermic reactions**. Ask students to consider whether or not all chemical reactions take place at the same speed. Does the rotting of an apple occur as quickly as the explosion of a piece of dynamite? Obviously, the answer is "No." Explain that some chemical reactions can be made to occur more quickly by adding a "catalyst." A **catalyst** is a substance that changes the speed of a chemical reaction without itself being changed by the reaction. Have students copy the definition of a <u>catalyst</u> on Journal Sheet #2.

Get Set!

Have students refer to Figure C on Journal Sheet #2. Have them copy the equation for the decompositon of hydrogen peroxide in the presence of a catalyst and go over the experiment described in Figure C.

$$\overset{\text{catalyst}}{\underset{\text{hydrogen peroxide}}{2H_2O_2}} \quad \rightarrow \quad \underset{\text{water}}{2H_2O} \quad + \quad \underset{\text{oxygen gas}}{O_2}$$

Explain that ferrous oxide, Fe_2O_3, or manganese (IV) dioxide, MnO_2, can be used as a catalyst to speed up the release of oxygen from hydrogen peroxide. Iron oxide is plain rust found on iron nails. Manganese dioxide is a brown solid that also acts as a catalyst in the production of poisonous chlorine gas from hydrochloric acid. Chlorine is an important gas used in the manufacture of bleach and other detergents (i.e., swimming pool cleansers).

Go!

Have students construct the set-up shown in Figure C on Journal Sheet #2 and complete the activity as directed.

Materials

beakers, test tubes, water, hydrogen peroxide, iron filings from rusted nails or manganese (IV) dioxide, hot plates, cardboards, scissors, wooden splints, matches

CH9 JOURNAL SHEET #2

TYPES OF CHEMICAL REACTIONS

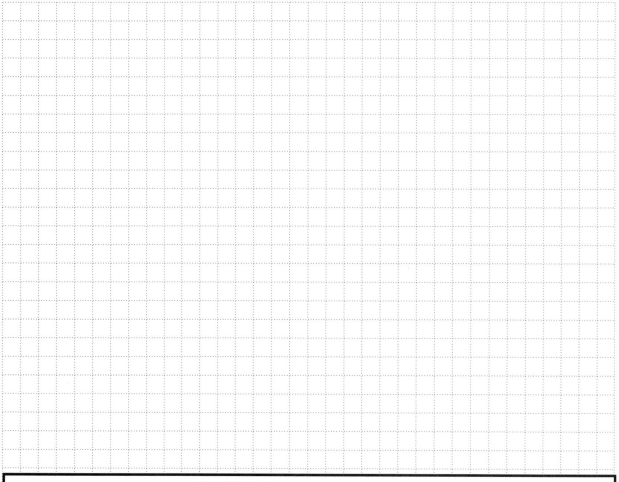

FIGURE C

Directions: (1) Cut a piece of cardboard to fit over the rim of a 500 ml beaker as shown below. (2) Label 4 test tubes with tape or pencil: Test tube #1, #2, #3, and #4. (3) Fill test tube #1 half way with water; test tube #2 half way with water and a pinch of either manganese dioxide or rusted iron filings; test tube #3 half way with hydrogen peroxide; and, test tube #4 half way with hydrogen peroxide and a pinch of manganese dioxide or rusted iron filings. (4) Fill the 500 ml beaker half way with water and place it on a hot plate. (5) Place the pieces of cardboard over the beaker and insert the test tubes as shown. (6) Turn the hot plate to high and heat the test tubes for 3-4 minutes. (7) Light a wooden splint with a match and allow it to burn for several seconds. (8) Gently blow out the flame so that the end of the splint glows red. (9) Introduce the "glowing splint" into each test tube one at a time. Relight the end of the splint if need be to keep the tip glowing. (10) Record your observations.

GENERAL SAFETY PRECAUTIONS

Be sure you are familiar with the proper use of the hot plate. Wear goggles to protect your skin and eyes from being burned by HOT WATER. Do not touch any part of the equipment without heat-resistant gloves or tongs. Clean up when the apparatus is cool.

top view of
beaker with
cardboard

"glowing
splint"

2 test
tubes in
front and
2 in back

cut
cardboard

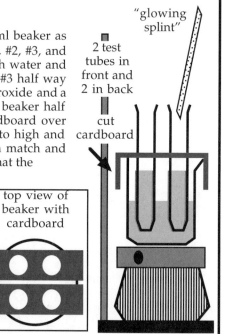

TYPES OF CHEMICAL REACTIONS

Work Date: ____/____/____

LESSON OBJECTIVE

Students will capture an explosive gas as the product of a single displacement reaction.

Classroom Activities

On Your Mark!

Working under a ventilation hood before the start of class, pour 100 ml of reagent grade hydrochloric acid (usually 37%) into a 250 ml beaker. Pour 900 ml of water into a 1,000 ml beaker or Ehrlenmeyer flask. Add the hydrochloric acid to the water (*not* vice versa). Cover or cap the solution and set it aside in a secure location. Prepare the one-holed rubber stopper-glass tube-rubber tubing assemblies (see Figure D on Journal Sheet #3).

In class, review the characteristics of a **single displacement reaction**. Write the general formula that appears on the Fact Sheet on the board and compare it to the chemical equation that describes the production of hydrogen gas by reaction of zinc and hydrochloric acid. Have students copy and compare the two chemical equations. If you have a short videotape of the *Hindenburg* or *Challenger* disaster, show it to emphasize the volatility of hydrogen gas!

$$X + YZ \rightarrow Y + XZ$$

$$Zn + 2HCl \rightarrow H_2 + ZnCl_2$$

Point out that the atoms of zinc "displace" the atoms of hydrogen in the acid to form the solid black product zinc chloride. During the reaction, hydrogen atoms pair up to form **diatomic molecules** that escape as a very light gas. In this experiment students will capture that gas by water displacement.

Get Set!

Spend ample time previewing the procedure described in Figure D on Journal Sheet #3. Be sure that students are aware of the "volatility" and "toxicity" of the substances used and produced in this demonstration. *Warn them to exercise extreme caution when dealing with the acid and gas.* Have students construct the set-up shown in Figure D and obtain their acid (about 20 ml per group) from you when they are ready to execute Step #6.

Go!

Assist students in performing the activity while exercising standard *laboratory safety guidelines*. For added effect, you can also ignite the gas accumulating in the Ehrlenmeyer flasks. *To minimize the chance of an accident do not allow students to do this!*

Materials

ring stand and clamps, 100 ml and 500 ml beakers, Ehrlenmeyer flasks, test tubes, rubber tubing, one-holed rubber stoppers, glass tubing, mossy zinc chips, hydrochloric acid, matches, cork stoppers

CH9 JOURNAL SHEET #3

TYPES OF CHEMICAL REACTIONS

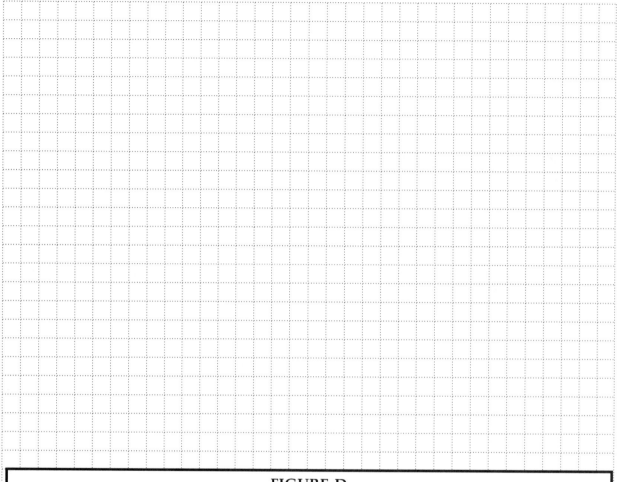

FIGURE D

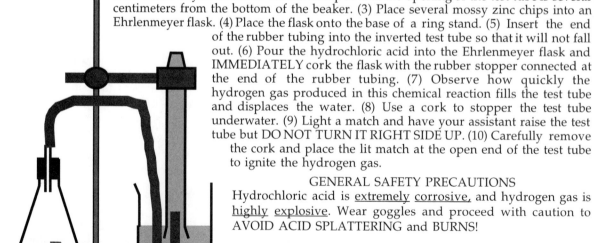

Directions: (1) Fill a 500 ml beaker halfway with water. (2) Fill a large test tube with water and using your thumb to prevent water from spilling out of the tube invert it into the beaker. Keep the tube filled with water as you secure it with a clamp so that the opening of the test tube is several centimeters from the bottom of the beaker. (3) Place several mossy zinc chips into an Ehrlenmeyer flask. (4) Place the flask onto the base of a ring stand. (5) Insert the end of the rubber tubing into the inverted test tube so that it will not fall out. (6) Pour the hydrochloric acid into the Ehrlenmeyer flask and IMMEDIATELY cork the flask with the rubber stopper connected at the end of the rubber tubing. (7) Observe how quickly the hydrogen gas produced in this chemical reaction fills the test tube and displaces the water. (8) Use a cork to stopper the test tube underwater. (9) Light a match and have your assistant raise the test tube but DO NOT TURN IT RIGHT SIDE UP. (10) Carefully remove the cork and place the lit match at the open end of the test tube to ignite the hydrogen gas.

GENERAL SAFETY PRECAUTIONS
Hydrochloric acid is extremely corrosive, and hydrogen gas is highly explosive. Wear goggles and proceed with caution to AVOID ACID SPLATTERING and BURNS!

TYPES OF CHEMICAL REACTIONS

Work Date: ____/____/____

LESSON OBJECTIVE

Students will form a precipitate as the product of a double displacement reaction.

Classroom Activities

On Your Mark!

A week before class, prepare a solution of iron acetate [$Fe(CH_3COO)_2$] by placing a teaspoon of rusted iron filings or steel wool in a 100 ml beaker of vinegar.

Begin class by performing the following demonstration: (1) Pour 20 ml of your iron acetate solution into another 100 ml beaker. (2) Add 20 ml of household ammonia: ammonium hydroxide [NH_4OH]. (3) Show students the green jellylike substance that forms. The products of this reaction are soluble ammonium acetate [$NH_4(CH_3COO)$] and insoluble ferrous hydroxide [$Fe(OH)_2$]. Point out that this insoluble substance is called a **precipitate**. The fer<u>rous</u> hydroxide will continue to absorb oxygen from the air to become a reddish-brown substance: fer<u>ric</u> hydroxide [$Fe(OH)_3$]. Explain that chemical reactions like this one, called "double displacement reactions," are used by chemists to create a variety of new and useful substances. Review the characteristics of a **double displacement reaction**. Write the general formula that appears on the Fact Sheet on the board. Compare it to the one that describes the reaction you just performed and the chemical equation that describes the production of lead iodide [PbI_2] and potassium nitrate [KNO_3] from potassium iodide [KI] and lead nitrate [$Pb(NO_3)_2$]. Both lead nitrate and potassium iodide are soluble white salts. The product potassium nitrate is a water soluble salt while lead iodide is a bright yellow, insoluble precipitate. Have students copy and compare the general equation to the two other chemical equations.

$$AB + CD \rightarrow AD + BC$$

$$Fe(CH_3COO)_2 + 2NH_4OH \rightarrow 2NH_4(CH_3COO) + Fe(OH)_2$$

$$2KI + Pb(NO_3)_2 \rightarrow 2KNO_3 + PbI_2$$

Get Set!

Spend ample time previewing the procedure described in Figure E on Journal Sheet #4. Be sure that students are aware of the "toxicity" of the substances used and produced in this demonstration. *Warn them to exercise extreme caution when dealing with materials containing high concentrations of lead.*

Go!

Have students perform the demonstration shown in Figure E on Journal Sheet #4 and complete the activity as directed. At the end of the demonstration pour the green/reddish-brown precipitate from your demonstration and the students' yellow mixture through a glass funnel into a 2-liter plastic bottle. Contact your district office to arrange for the proper disposal of the material. *Do not pour it down the sink.* Lead iodide is a toxic waste used in the production of lead-based paints.

Materials

plastic vials, 100 ml beakers, "spatulas" made of plastic straws cut lengthwise, vinegar, steel wool or rusted iron filings, household ammonia or ammonium hydroxide, lead nitrate, potassium iodide, glass funnel, 2-liter plastic bottle

CH9 Journal Sheet #4

TYPES OF CHEMICAL REACTIONS

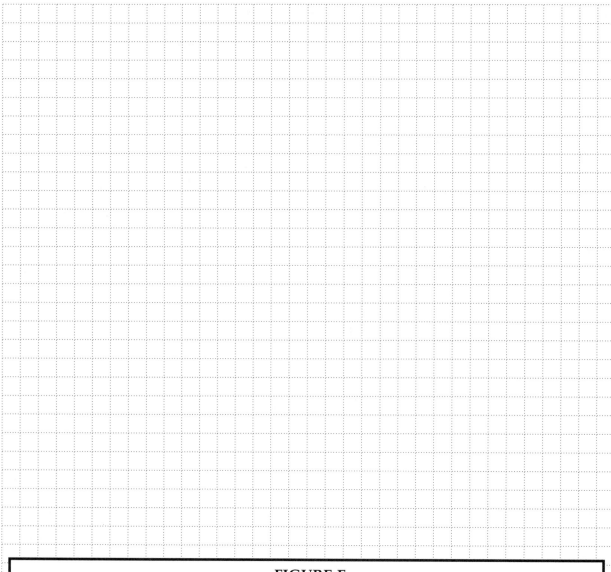

FIGURE E

Directions: (1) Fill 2 small 100 ml beakers with 20 ml of water. (2) Using the "straw spatula" provided by your instructor, add a small scoop of potassium io-dide (KI) to one of the beakers and stir. (3) Using a second "straw spatula" add a scoop of lead nitrate ($Pb(NO_3)_2$) to the

vial of potassium iodide (KI)

solution of potassium iodide (KI)

mixed solution of lead iodide and potassium nitrate ($PbI_2 + KNO_3$)

solution of lead nitrate $Pb(NO_3)_2$

vial of lead nitrate $Pb(NO_3)_2$

other beaker and stir. (4) On the command of your instructor, pour the contents of the two beakers into a third empty beaker and record your observations.

GENERAL SAFETY PRECAUTIONS

The salts used and produced in this experiment are <u>toxic</u>. AVOID CONTACT WITH SKIN and follow the directions of your instructor in disposing of all materials.

CH9 REVIEW QUIZ

Directions: Keep your eyes on your own work.
Read all directions and questions carefully.
THINK BEFORE YOU ANSWER!
Watch your spelling, be neat, and do the best you can.

CLASSWORK   (~40): _____
HOMEWORK   (~20): _____
CURRENT EVENT   (~10): _____
TEST   (~30): _____

TOTAL   (~100): _____
(A ≥ 90, B ≥ 80, C ≥ 70, D ≥ 60, F < 60)

LETTER GRADE: _____

TEACHER'S COMMENTS: _____
_____

TYPES OF CHEMICAL REACTIONS

TRUE–FALSE FILL-IN: If the statement is true, write the word TRUE. If the statement is false, change the underlined word to make the statement true. *15 points*

_____ 1. Chemists use symbolic sentences called chemical <u>equations</u> to describe how elements and compounds behave when mixed.

_____ 2. The substances mixed together in a chemical reaction are called the <u>products</u>.

_____ 3. The substances produced in a chemical reaction are called the <u>reactants</u>.

_____ 4. In a <u>synthesis</u> reaction individual substances combine to form a single new substance.

_____ 5. A <u>cofactor</u> tells the chemist how many molecules of a substance are involved in any chemical reaction.

_____ 6. The synthesis of water is an <u>exothermic</u> reaction that releases an enormous amount of energy.

_____ 7. An <u>endothermic</u> chemical reaction was responsible for the explosive destruction of the giant airship *Hindenburg* in 1937 and the space shuttle *Challenger* in 1986.

_____ 8. A compound is split into smaller chemical units in a <u>single displacement</u> reaction.

_____ 9. The electrolysis of water to form oxygen and hydrogen gas is a <u>single displacement</u> reaction.

_____ 10. The electrolysis of water is an <u>endothermic</u> reaction because electrical energy is absorbed by water molecules to produce individual oxygen and hydrogen molecules.

_____ 11. The production of hydrogen gas during the mixing of an acid with a metal is a <u>single displacement</u> reaction.

_____ 12. In a <u>double displacement</u> reaction, atoms in different compounds "displace" one another to form new compounds.

_____13. The acid-base reaction is the most commonly used <u>synthesis</u> reaction.

_____14. A <u>double displacement</u> reaction takes place in your stomach when you eat an antacid to cure an upset stomach.

_____15. Laplace and Lavoisier demonstrated that water is a(n) <u>element</u>.

PROBLEM

MATCHING: Choose the letter of the chemical reaction that best matches the class to which it belongs. *12 points*

_____ 16. synthesis

_____ 17. decomposition

_____ 18. single displacement

_____ 19. double displacement

(A) $AB + C \rightarrow BC + A$

(B) $M + N \rightarrow MN$

(C) $QR + ST \rightarrow QS + RT$

(D) $WXYZ \rightarrow WX + YZ$

Chemical Equation: Chemical engineers are already designing automobile engines for the 21st century that will be able to run on water. Write a balanced chemical equation for the decomposition of water that would produce hydrogen and oxygen gas to use as fuel. *3 points*

_____   _____   ____/____/____
Student's Signature   Parent's Signature   Date

ACID-BASE REACTIONS

TEACHER'S CLASSWORK AGENDA AND CONTENT NOTES

Classwork Agenda for the Week

1. Students will list the physical and chemical properties of acids and bases.
2. Students will use litmus paper to measure the pH of acids and bases.
3. Students will demonstrate that acids and bases form ionic solutions.
4. Students will neutralize a base by titration with an acid.

Content Notes for Lecture and Discussion

Medieval alchemists were familiar with the corrosive properties of sulfuric acid—which they called "oil of vitriol"—and a mixture of hydrochloric and nitric acids—which they called "aqua regia." Both were used as powerful solvents in the manufacture of a variety of textile products. But it was not until the early 17th century that chemists improved methods for preparing and purifying acids. The German chemist **Johann Rudolf Glauber** (b. 1604; d. 1670) manufactured pure, concentrated hydrochloric acid by distilling a mixture of sulfuric acid and common table salt. The acid was used commercially to prepare a number of useful salts from metals and metal oxides. The Irish chemist **Robert Boyle** (b. 1627; d. 1691) catalogued the bitter and corrosive properties of acids and theorized that their properties were the result of the "pointed" nature of the particles that comprised them. Boyle's hypothesis was replaced by the notion of "attractive and repulsive forces" suggested by the work of **Sir Isaac Newton** (b. 1642; d. 1727). After Newton, the neutralization of an acid by a base (or vice versa) could be explained in terms of the neutralization or "cancelling" of these forces.

The French chemist **Antoine Laurent Lavoisier** (b. 1743; d. 1794) proposed that acidity was a property of oxygen-containing compounds. The word oxygen means "acid producer." And, Lavoisier's work with sulphates and phosphates, both oxygen-containing molecules, supported his theory. However, the English chemist **Humphrey Davy** (b. 1778; d. 1829), who employed electrolysis to discover the elements boron, barium, and calcium among others, proved Lavoisier wrong. He showed that oxygen was in fact a component of bases like the alkalies sodium and potassium hydroxide. He also demonstrated that there was no oxygen present in hydrochloric acid. Although he never proposed a sufficient theory to explain the behavior of acids, Davy suspected that hydrogen was the key element involved in the phenomenon of acidity. By the mid-19th century, scientists using the electrolysis methods popularized by Davy began characterizing acidic solutions by the amount of "displaceable" hydrogen in a solution.

The **theory of ionic dissociation** proposed by the Swedish chemist **Svante August Arrhenius** (b. 1859; d. 1927) became the basis of a new theory of acids and bases. According to Arrhenius' theory, ionic dissociation is a reversible process and the **electrolytes** (i.e., charged particles) in a solution reach equilibrium, establishing a balance between the dissociated and undissociated molecules in the solution. The degree of dissociation can be measured and is specific for each compound. Arrhenius proposed that ions could be considered "independent" molecular particles having their own special physical and chemical characteristics. This notion is evidenced by the fact that sodium atoms have very different properties than sodium ions. Ionic dissociation explains how acids "donate" hydrogen to solutions. *Acids are hydrogen donors. Bases are hydrogen acceptors.*

CH10 Content Notes *(cont'd)*

In 1909, the Danish chemist **Søren Peter Lauritz Sørensen** (b. 1868; d. 1939) proposed that strong acids contained hydrogen ions in concentrations of about 1 gram of ion per liter of solution. Bases contained as little as 10^{-14} grams of dissociated hydrogen ion in one liter of solution. Sørensen defined the pH of a solution as the negative logarithm of hydrogen ion concentration (i.e., -log [H$^+$]). For example, a solution containing a 10^{-5} molar concentration of hydrogen ions has a pH equal to 5. Recall that log 100 = 2 (in base "10") because $10^2 = 100$. Strong acid used in car batteries have a pH of about 2; citrus fruits have a pH of about 4; soil has a pH of about 7; soaps have a pH of about 10; and, potash (potassium hydroxide) has a pH of about 13.

In Lesson #1, students will list the physical and chemical properties of acids and bases and prepare a 0.1 molar solution of a base.

In Lesson #2, students will use litmus paper to measure the pH of acids and bases.

In Lesson #3, students will make discoveries similar to those of Humphrey Davy and Søren Sørensen and demonstrate that acids and bases form ionic solutions.

In Lesson #4, students will "stain" a base using phenophthalein indicator and neutralize the base by titration with an acid.

ANSWERS TO THE HOMEWORK PROBLEMS

Tables of household acids and bases will vary. Be sure students identify the presence of an acid by recognizing the term "acid" in the ingredients of listed products. Be sure students identify the presence of a base by recognizing the term "hydroxide" in the ingredients of listed products.

ANSWERS TO THE END-OF-THE-WEEK REVIEW QUIZ

1. acids
2. true
3. and
4. hydrogen
5. bases

6. hydroxide
7. hydroxide
8. true
9. true
10. water

11. acid
12. acid
13. salt
14. base
15. acid

16. base
17. salt
18. acid
19. base
20. base

Unaware of the proper rules for writing chemical formulas students may not place symbols in their correct order. Excuse these mistakes but emphasize the proper use of subscripts and balancing coefficients.

21. $HCl + NaOH \rightarrow NaCl + H_2O$
22. $HNO_3 + NaOH \rightarrow NaNO_3 + H_2O$
23. $H_2SO_4 + 2NH_4OH \rightarrow (NH_4)_2SO_4 + 2H_2O$
24. $H_2CO_3 + 2RbOH \rightarrow Rb_2CO_3 + 2H_2O$
25. $2HCl + Mg(OH)_2 \rightarrow MgCl_2 + 2H_2O$

CH10 Fact Sheet

ACID-BASE REACTIONS

CLASSWORK AGENDA FOR THE WEEK

(1) List the physical and chemical properties of acids and bases and prepare a 0.1 molar solution of a base.
(2) Use litmus paper to measure the pH of acids and bases.
(3) Demonstrate that acids and bases form ionic solutions.
(4) Neutralize a base by titration with an acid.

Acids are one of the most common groups of chemical substances. Acids are sour tasting, extremely corrosive, good conductors of electricity, and will turn **litmus paper** red. They also react with metals, sometimes violently, to liberate explosive hydrogen gas. All acids contain a **hydrogen ion (H^+)** that goes into solution when mixed with water. The hydrogen ion (a single positive proton) can "tear" electrons from other chemical substances. This accounts for an acid's "biting" physical and "reactive" chemical properties. Notice the similarities in the chemical formulas of the following common acids:

hydrochloric acid	HCl
sulfuric acid	H_2SO_4
carbonic acid	H_2CO_3
nitric acid	HNO_3

Bases are also very common chemical substances. Bases are bitter tasting, extremely caustic, good conductors of electricity, and will turn litmus paper blue. All bases contain a **hydroxide ion (OH^-)** that goes into solution when mixed with water. Because of its negative charge, the hydroxide ion attracts positive ions of any kind. This accounts for the "irritating" physical and "reactive" chemical properties of bases. Bases can **neutralize** acids to form water and salt. Notice the similarities in the chemical formulas of the following list of common bases:

sodium hydroxide	$NaOH$
magnesium hydroxide	$Mg(OH)_2$
ammonium hydroxide	NH_4OH
potassium hydroxide	KOH

Neutralizing an acid with a base is called a **neutralization reaction**. The general chemical equation for this type of chemical reaction can be expressed as follows:

$$\textbf{HX} + \textbf{YOH} \rightarrow \textbf{HOH} + \textbf{XY}$$

acid	base	water	salt

A neutralization reaction is a **double displacement** reaction.

In an "acid-base," or neutralization reaction, two "harsh" substances are transformed to harmless **water** and **salt**. Since the manufacture of many kinds of materials involves the use of acids and bases, the neutralization reaction is one of the most common chemical reactions performed by modern industry.

Homework Directions

1. With the permission of your parent/guardian, read the ingredients labels of five (5) common household substances. Create a table that lists those five (5) common substances and the acids they contain. The term "acid" is used to identify a substance as an acid. *7 points*

2. With the permission of your parent/guardian, read the ingredients labels of five (5) common household substances. Create a table that lists those five (5) common substances and the bases they contain. The term "hydroxide" is used to identify a substance as a base. *8 points*

Assignment due: _____

_____   _____   ____/____/____
Student's Signature   Parent's Signature   Date

CH10 Lesson #1

ACID-BASE REACTIONS

Work Date: ____/____/____

LESSON OBJECTIVE

Students will list the physical and chemical properties of acids and bases.

Classroom Activities

On Your Mark!

Display a lemon, a bottle of household cleaner containing ammonium hydroxide, a bottle of vinegar, a bar of soap, and a roll of antacid tablets. Point out that both the lemon and the vinegar contain an **acid** called "citric acid." The presence of citric acid gives the **family of citrus fruits** their name. Another acid found in citrus fruits is "ascorbic acid," commonly called "vitamin C." Point out that acids are common in our everyday lives. We even have acid in our stomach to help us digest the foods we eat: a powerful solution of "hydrochloric acid." Point out that the cleaner, the soap, and the antacid tablets contain substances called **bases**. Bases have a variety of uses and are most commonly employed to "neutralize" acids used in a myriad of industrial and chemical procedures. Ask students to compare the chemical formulas of the acids and bases listed on the Fact Sheet. Write several of these formulas on the board. Ask: How are the acid formulas the same? Answer: each formula has an "H" for **hydrogen** at the front of it. All acids contribute a **hydrogen ion** (i.e., H^+) to water when in solution. Ask: How are the base formulas the same? Answer: each formula has an **OH** for **hydroxide** at the back of it. All bases contribute a **hydroxide ion** (i.e., OH^-) to water when in solution.

Get Set!

Compare and contrast the physical and chemical properties of acids and bases by drawing Illustration A on the board. Have students copy these notes on Journal Sheet #1. Ask a student volunteer to bite into the lemon—or do it yourself—to demonstrate the meaning of the term "sour." Something that tastes "sour" makes us "pucker our lips." Ask students what

> ### ILLUSTRATION A
> #### PHYSICAL AND CHEMICAL PROPERTIES OF ACIDS AND BASES
>
> | ACIDS | BASES |
> |---|---|
> | *taste sour | *taste bitter |
> | *corrosive (i.e., causes disintegration) | *caustic (i.e., causes a burning sensation) |
> | *form ionic solutions that conduct electricity | *form ionic solutions that conduct electricity |
> | *turn litmus paper red | *turn litmus paper blue |

they would be inclined to do if they bit into a bar of soap. Soap has a "bitter" taste that makes one want to expectorate. Take out a piece of **litmus paper** and demonstrate how it changes color when touched to the lemon or dipped into the household cleaner. Explain that **litmus** is a dye obtained from lichen: a symbiotic community of algae and fungi that grow on rocks. The dye is sensitive to the presense of acids and bases. Litmus turns red in the presence of an acid and blue in the presence of a base. Explain that students will test for the presence of acids and bases in Lesson #2 and neutralize a base using an acid in Lesson #4. In order to do this, they will learn how to prepare and use a 0.1 molar solution of a common base—sodium hydroxide (NaOH).

Go!

Assist students in preparing a 0.1 molar solution of sodium hydroxide as directed in Figure A on Journal Sheet #1. The atomic masses of Na, O, and H are 23, 16, and 1, respectively. One mole of NaOH is, therefore, 40 grams of NaOH. One twentieth or 0.05 of 40 is 2 grams. Mixing 2 grams of NaOH in 500 ml of water will yield a 0.1 molar solution of that base.

Materials

a lemon, a bottle of household cleaner containing ammonium hydroxide, a bottle of vinegar, a bar of soap, a roll of antacid tablets, sodium hydroxide pellets, hot plates, balances, glass stirring rods, 500 ml beakers or Ehrlenmeyer flasks, corks (or wax paper and rubber bands) for flasks

CH10 JOURNAL SHEET #1

ACID-BASE REACTIONS

TABLE A

chemical symbol or formula	atomic mass
Na	_____
O	_____
H	_____
NaOH	_____
	÷ 20
	= 0.05 mole

to mix _____ grams NaOH in 500 ml water to get a 1 molar solution.

FIGURE A
Preparation of a 0.1 molar solution of sodium hydroxide

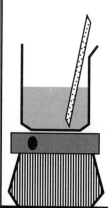

Directions: (1) Use THE PERIODIC TABLE OF THE ELEMENTS to find the atomic mass of sodium (Na) and enter that mass on Table A. Find the masses of oxygen (O) and hydrogen (H) as well. The atomic mass of these elements is equal to one mole (or Avogadro's number = 6×10^{23} atoms) of that element. Add the masses together to get the mass of 1 mole of sodium hydroxide (NaOH). (2) Use a balance to prepare one twentieth (0.05) of a mole of NaOH. (3) Fill a 500 ml beaker or 500 ml Ehrlenmeyer flask with water and place it on a hot plate. (4) Turn the hot plate on a medium setting and add the NaOH. Since a one molar solution of any substance is the molecular weight of that substance in 1,000 ml of water, the mixture will be 0.1 molar solution of NaOH. (5) Stir gently until all the NAOH is completely dissolved in the solution. Use wax paper and a rubber band, or a cork or rubber stopper, to cover or cap the solution. (6) Store the solution at room temperature for use in Lesson #2.

GENERAL SAFETY PRECAUTIONS

Be sure you are familiar with the proper use of the hot plate. Wear goggles to protect your skin and eyes from being burned by HOT WATER. Avoid contact with the sodium hydroxide pellets. They are caustic. Do not touch any part of the equipment without heat-resistant gloves or tongs. Clean up only when the apparatus is cool.

CH10 Lesson #2

ACID-BASE REACTIONS

Work Date: ____/____/____

LESSON OBJECTIVE

Students will use litmus paper to measure the pH of acids and bases.

Classroom Activities

On Your Mark!

Before the start of class prepare a 0.1 molar solution of HCl from bottled acid obtained from a laboratory supply house. If you purchase bottles of 1 molar HCl dilute the solution 10:1 with water. If you use concentrated HCl (usually bottled in 37% solutions) you can dilute it as follows: A one molar hydrochloric acid solution contains 17 grams of HCl in 1,000 ml of water. A 1,000 ml 37% HCl solution contains 370 grams of HCl and 630 ml water. This is a 34.5 molar solution of HCl—a powerful acid! To obtain a 0.1 molar solution from this concentrated solution, dilute it approximately 300:1 (i.e., add 900 ml of water to about 30 ml of the concentrated acid).

Review the physical and chemical properties of acids and bases. Inform students that **litmus** is a mixture of several **organic compounds** (i.e., compounds containing mostly carbon, hydrogen, and oxygen) that change color in the presence of an acid or base. Litmus can be extracted from species of **lichen** grown in the Netherlands: *Lecanora tartarea* and *Roccella tinctorum*. The Irish chemist **Robert Boyle** (b. 1627; d. 1691) was one of the first to organize the analysis of acid, base, and neutral solutions using plant extracts as **indicators**. Boyle published his methods in 1664 in a book entitled *Experiments and Considerations Touching Colours*. Litmus paper can be obtained from laboratory supply houses in strips or rolls. For the purposes of Lesson #2, a roll of litmus paper that is sensitive to a range of acid and base solutions is most ideal. Most rolls come complete with a "color legend" that matches colors to a range of **pH** (i.e., concentration of H^+ and OH^- **ions**).

Get Set!

Draw the **pH SCALE** shown to the right and compare it to the color legend on the litmus paper roll. Explain that the color created by a drop of acid or base on the litmus paper reflects the concentration of hydrogen or hydroxide ions present in solution. More advanced students can learn that pH is the negative logarithm of hydrogen ion concentration as explained in the Teacher's Agenda and Content Notes.

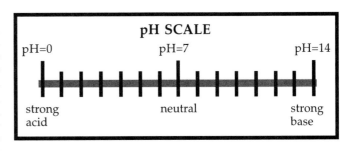

pH SCALE

pH=0   pH=7   pH=14

strong acid   neutral   strong base

Go!

Assist students in performing the experiment described in Figure B. The experiment in Figure B will allow students to make qualitative judgements regarding the pH of a variety of acids and bases.

Materials

a lemon, a bottle of household cleaner containing ammonium hydroxide, a bottle of vinegar, a bar of soap, a roll of antacid tablets, sodium hydroxide solution prepared in Lesson #1, mild hydrochloric acid solution, medicine droppers, litmus paper rolls, water, small beakers

Name: _____ Period: _____ Date: ___/___/___

CH10 JOURNAL SHEET #2

ACID-BASE REACTIONS

FIGURE B

Directions: (1) Place strips of litmus paper in the spaces provided (Test A, Test B, etc). (2) Test single-drop samples of the solutions given to you by your instructor using a medicine dropper (lemon juice, household cleaner, 0.1 molar solution of sodium hydroxide, 0.1 molar solution of hydrochloric acid, etc.). (3) Compare the color created by the sample drop against the "color legend" on the litmus paper dispenser. Record the approximate pH of each solution. (4) Use clear tape to secure the samples to the page making sure to label each solution used.

GENERAL SAFETY PRECAUTIONS

Wear goggles to protect your eyes against splatterered fluids and avoid getting the fluids on your clothing.

Test A	Solution: _____ /pH: _____
Test B	Solution: _____ /pH: _____
Test C	Solution: _____ /pH: _____
Test D	Solution: _____ /pH: _____
Test E	Solution: _____ /pH: _____
Test F	Solution: _____ /pH: _____

ACID-BASE REACTIONS

Work Date: ____/____/____

LESSON OBJECTIVE

Students will demonstrate that acids and bases form ionic solutions.

Classroom Activities

On Your Mark!

Begin class discussion with a review of the definition of a **solution**. A solution is a liquid mixture containing a **solute** and a **solvent**. A solute is the substance that is dissolved in a liquid. A solvent is the liquid in which a substance is dissolved. Ask students to discuss the characteristics of a good **conductor** of electricity. A good conductor is one that allows the free flow of electrical charges. **Electrons** flow freely through a metal conductor. **Ions** flow freely in a liquid solvent. Remind students that ions are charged particles formed when atoms lose or gain electrons.

Get Set!

Explain that ionic solutions are good conductors of electricity. Write the formulas of the following acids and bases and explain to students that both acids and bases form **ions** in water solution.

dissociation of an acid	dissociation of a base
$HCl \rightarrow H^+ + Cl^-$	$NaOH \rightarrow Na^+ + OH^-$

Have students copy this information on Journal Sheet #3.

Go!

Assist students in performing the experiment described in Figure C. Have them test the water alone to discover that the bulb does not light on the power of a single D-cell battery. Addition of several drops of lemon juice, household cleaner, the 0.1 molar solution of NaOH they prepared in Lesson #1, or a drop of the 0.1 molar solution of HCl they used in Lesson #2 will light the bulb to varying degrees of brightness. The experiment in Figure C will demonstrate that water is not a good conductor of electricity unless it contains ions. They will see that ionic solutions containing either an acid or a base are good conducting fluids; therefore, they must be ionic solutions.

Materials

a lemon, a bottle of household cleaner containing ammonium hydroxide, a bottle of vinegar, a bar of soap, a roll of antacid tablets, sodium hydroxide solution prepared in Lesson #1, mild hydrochloric acid solution (i.e., 0.1 molar solution), medicine droppers, water, flashlight bulb and socket, D-cell battery, fine gauge insulated wire, switch, water, glass stirring rods

CH10 JOURNAL SHEET #3

ACID-BASE REACTIONS

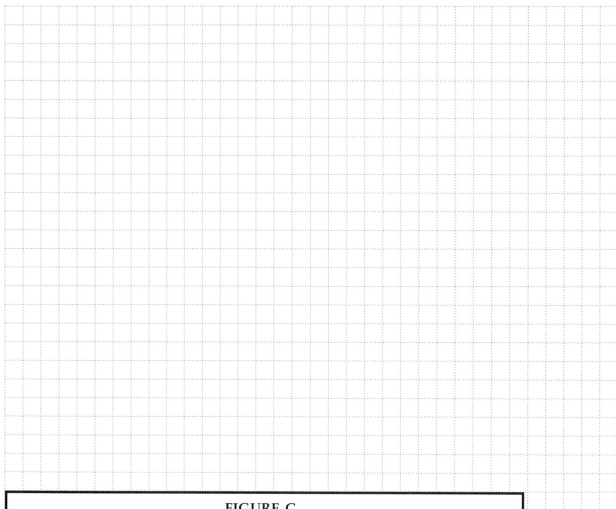

FIGURE C

Directions: (1) Fill a 100 ml beaker with water. (2) Construct the set-up shown below. (3) Close the switch and test to see if the bulb lights. Record your observations. (4) Test the solutions given to you by your instructor one at a time by opening the switch and rinsing the beaker after each test (i.e., lemon juice, household cleaner, 0.1 molar solution of sodium hydroxide, 0.1 molar solution of hydrochloric acid, etc.). (5) Summarize your conclusions about the "conductivity" of acids and bases by citing test observations in your explanation .

GENERAL SAFETY PRECAUTIONS

As each solution begins to conduct electricity, bubbles of TOXIC GAS will form at the electrodes
and escape. DO NOT INHALE THESE GASES. Wear goggles to protect your eyes against splattered fluids and DO NOT TOUCH THE SOLUTION WHILE THE SWITCH IS CLOSED.

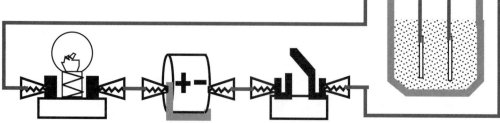

CH10 Lesson #4

ACID-BASE REACTIONS

Work Date: ____/____/____

LESSON OBJECTIVE

Students will neutralize a base by titration with an acid.

Classroom Activities

On Your Mark!

Phenolphthalein indicator will be used as a base indicator in this lesson and is available in solution through any laboratory supply house. Phenolphthalein is clear below pH 7 and purple above pH 7. When titrating a base with an acid, it is convenient to "stain" the base with phenolphthalein so that the neutralization of the base is evident when the solution clears.

Begin discussion by asking students what they commonly do when they have an upset stomach. Explain that "antacids" contain **hydroxides** that neutralize acids (such as HCl in stomach acid) according to the following general chemical equation:

$$HX \quad + \quad YOH \quad \rightarrow \quad XY \quad + \quad HOH$$
$$\text{acid} \qquad \text{base} \qquad \quad \text{salt} \qquad \text{water}$$

Draw Chart A one row at a time to show how a variety of common acids and bases are **neutralized** to form salt and water.

CHART A			
acid +	base ⟶	salt +	water
HCl stomach acid	$NaOH$ lye for soap	$NaCl$ table salt	HOH H_2O = water
H_2CO_3 soda pop acid	$NaOH$ lye for soap	Na_2CO_3 washing soda	HOH H_2O = water
H_2SO_4 industrial acid	$Mg(OH)_2$ milk of magnesia	$MgSO_4$ Epsom salt	HOH H_2O = water
H_2CO_3 soda pop acid	$Ca(OH)_2$ antacid tablets	$CaCO_3$ limestone chalk	HOH H_2O = water
HNO_3 fertilizer acid	KOH potash	KNO_3 saltpeter	HOH H_2O = water

Get Set!

Prepare a 100 ml beaker of 0.1 molar NaOH and another 100 ml of 0.1 molar (or slightly stronger) HCl before the start of class. Put a drop of phenolphthalein into an empty beaker when students are not looking. Add some NaOH solution to the beaker with the indicator. Watching the clear liquid turn purple as you pour it into the "empty" beaker will raise some questions. Pour a few drops of the purple fluid into the acid which will remain clear. This will also raise eyebrows. Pour the acid into the base until the base clears. Identify all of your solutions and explain that phenolphthalein can only indicate the presence of a base when there is base present. When the acid neutralized the base to form salt and water, the phenolphthalein became colorless.

Go!

Assist students in performing the experiment described in Figure D on Journal Sheet #4.

Materials

beakers, test tubes or 10 ml graduated cylinders, 0.1 molar NaOH, 0.1 molar HCl, rubber or cork stoppers, medicine droppers

CH10 JOURNAL SHEET #4

ACID-BASE REACTIONS

FIGURE D

Directions: (1) Use a test tube rack or cut a hole in a piece of cardboard to hold a test tube upright in a beaker. If a 10 ml graduated cylinder is available then neither the test tube rack nor test tubes are necessary. Use the graduated cylinder. (2) Set aside 20 ml of the 0.1 molar solution of sodium hydroxide you prepared in Lesson #1 in a 100 ml beaker. (3) Obtain 20 ml of a 0.1 molar solution of hydrochloric acid from your instructor and set it aside in a 100 ml beaker. (4) Pour 3 centimeters of your 0.1 molar sodium hydroxide solution into the test tube (or 3 ml into the graduated cylinder). (5) Add one drop of the indicator solution (i.e., phenolphthalein) to the sodium hydroxide. The solution will turn purple indicating the presence of a base. (6) Measure the height of the solution using a ruler.

(7) Add a dropper full of the hydrochloric acid solution to the base. (8) Use a small rubber or cork stopper to cap the tube or cylinder and turn it upside down then right side up. This simple action will mix the solution thoroughly. (9) Repeat steps #7 and #8 until the base solution "clears" completely. (10) Use a ruler to measure the amount of acid it took to neutralize the base. NOTE: You can also perform this experiment with a 10 ml graduated cylinder. This will allow you to accurately measure the amount of acid needed to neutralize the base.

GENERAL SAFETY PRECAUTIONS

Wear goggles to protect your eyes against splattered fluids and avoid getting the solution on clothing.

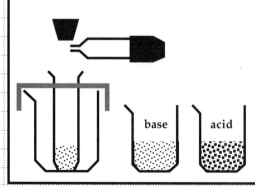

base   acid

CH10 Review Quiz

Directions: Keep your eyes on your own work.
Read all directions and questions carefully.
THINK BEFORE YOU ANSWER!
Watch your spelling, be neat, and do the best you can.

CLASSWORK   (~40): _____
HOMEWORK   (~20): _____
CURRENT EVENT (~10): _____
TEST   (~30): _____

TOTAL   (~100): _____
(A ≥ 90, B ≥ 80, C ≥ 70, D ≥ 60, F < 60)

LETTER GRADE: _____

TEACHER'S COMMENTS: _____

_____

ACID-BASE REACTIONS

TRUE–FALSE FILL-IN: If the statement is true, write the word TRUE. If the statement is false, change the underlined word to make the statement true. *10 points*

_____ 1. <u>Bases</u> react with metals to produce hydrogen gas.

_____ 2. Acids contain a(n) <u>hydrogen</u> ion that goes into solution when mixed with water.

_____ 3. Acids <u>but not</u> bases conduct electricity.

_____ 4. The <u>hydroxide</u> ion is a single positive proton.

_____ 5. Bitter tasting <u>acids</u> are sometimes used to make soap.

_____ 6. Bases contain a(n) <u>hydrogen</u> ion that goes into solution when mixed with water.

_____ 7. A(n) <u>hydrogen</u> ion contains hydrogen and oxygen.

_____ 8. The <u>products</u> of a neutralization reaction are water and salt.

_____ 9. The <u>reactants</u> of a neutralization reaction are an acid and a base.

_____10. In a neutralization reaction hydrogen and hydroxide ions combine to produce <u>salt</u>.

Write the word ACID, BASE, or SALT next to each chemical formula. *10 points*

_____ 11. HCl

_____ 12. HNO_3

_____ 13. KI

_____ 14. NH_4OH

_____ 15. H_2CO_3

_____ 16. $NaOH$

_____ 17. $CaCl_2$

_____ 18. H_2SO_4

_____ 19. $RbOH$

_____ 20. NH_4OH

PROBLEM

Directions: Fill in the chemical formulas of the missing reactants and products in each chemical reaction. Fill in the coefficients to balance the equations. *10 points*

21. ____ HCl + ____ NaOH → ____ NaCl + ____ _____

22. ____ _____ + ____ NaOH → ____ NaNO$_3$ + ____ H$_2$O

23. ____ H$_2$SO$_4$ + ____ NH$_4$OH → ____ (NH$_4$)$_2$SO$_4$ + ____ _____

24. ____ H$_2$CO$_3$ + ____ RbOH → ____ _____ + 2H$_2$O

25. ____ HCl + ____ Mg(OH)$_2$ → ____ _____ + 2H$_2$O

_____   _____   ____/____/____
Student's Signature   Parent's Signature   Date

CARBON COMPOUNDS
AND PETROCHEMICALS

TEACHER'S CLASSWORK AGENDA AND CONTENT NOTES

Classwork Agenda for the Week

1. Students will draw structural formulas and build models of carbon compounds.

2. Students will discover the kinds of rocks that serve as crude oil reservoirs and draw a diagram to show how petroleum is refined.

3. Students will determine the number of calories in a piece of charcoal.

4. Students will break up hydrocarbon molecules by catalytic cracking.

Content Notes for Lecture and Discussion

Wood was the first organic fuel used for cooking. Charcoal was used for smelting iron ores as early as 5,000 B.C. in both Europe and Mesopotamia; and, the Romans used coal as a fuel in the 1st century A.D. In 1709, English iron manufacturer **Abraham Darby** (b. 1677; d. 1717) developed the first process for smelting iron using **coke**: a hotter burning, more energy efficient fuel than charcoal. **Bitumen**—a form of thick oil that seeps from the ground over shallow oil deposits—was used by ancient Mesopotamian civilizations as a tar to seal the hulls of sailing vessels. But, it was not until the first decade of the 19th century that chemists made a distinction between **organic** and **inorganic** substances. The Swedish chemist **Jöns Jakob Berzelius** (b. 1779; d. 1848) suggested that the products of living organisms, like sugar and vegetable oils, should be called "organic." He used the term "inorganic" to describe substances that did not originate from living matter such as sulphate salts and metal oxides. Before the German chemist **Friedrich Wöhler** (b. 1800; d. 1882) succeeded in synthesizing urea (an organic substance) from ammonium cyanate (an inorganic substance) in 1828, the synthesis of organic materials from inorganic ones was considered impossible. It was believed that a "vital force" was necessary to create "living molecules" from "nonliving molecules." This **theory of vitalism** was dashed in the following decades as chemists succeeded in synthesizing a host of organic compounds from inorganic ones.

Organic chemistry had its origin with the work of the German chemist **Friedrich August Kekúle von Stradonitz** (b. 1829; d. 1896). Kekúle worked out the structure of benzene (1,3,5 cyclohexene = C_6H_6) and revolutionized chemists' conceptualization of molecular structure by proposing the three-dimensional tetrahedral structure of methane. Kekúle's work laid the foundation for modern **synthetic organic chemistry**. Following the drilling of the first oil well in Titusville, Pennsylvania, by **Edwin Laurentine Drake** (b. 1819; d. 1880) in 1859, crude oil became a staple fuel and source of organic chemicals. In 1862, English chemist **Alexander Parkes** (b. 1813; d. 1890) synthesized the first **thermoplastic** from the plant extract cellulose

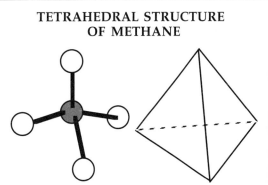

TETRAHEDRAL STRUCTURE OF METHANE

The methane molecule contains four hydrogen atoms bonded covalently to a central carbon. The hydrogen atoms are equidistant from one another: a result of the mutual repulsion between the electron pairs constituting the "single" bonds.

nitrate. The French chemist **Hilaire Bernigaud Chardonnet** (b. 1839; d. 1924), produced the first artifical fiber—rayon—in 1884. And, Belgian chemist **Leo Hendrik Baekeland** (b. 1863; d. 1944) synthesized the first commercial plastic, Bakelite, by reaction of formaldehyde and phenol. The American chemist **Wallace Hume Carothers** (b. 1896; d. 1937) mastered the science of polymerization to produce the first strands of nylon fiber in 1935. The pharmaceutical industry began with the work of German chemists **Richard Willstätter** (b. 1872; d. 1942) and **Heinrich Otto Wieland** (b. 1877; d. 1957) who worked out the structure of chlorophyll and steroids, respectively. The German bacteriologist **Paul Ehrlich** (b. 1854; d. 1915) used synthetic drugs to treat disease, thereby introducing the science of **chemotherapy** to the field of medicine.

In Lesson #1 students will be introduced to the versatility and enormous variety of carbon compounds. They will construct and draw models and structural formulas for families of carbon compounds: alkanes, alkenes, and alkynes.

In Lesson #2 students will be introduced to the methods used to find and refine crude oil and discover the kinds of rock that serve as crude oil reservoirs.

In Lesson #3 students will discover the considerable amount of energy stored in fossil fuels by determining the number of calories in a piece of charcoal.

In Lesson #4 students will use catalytic cracking—a common chemical engineering technique—to convert large hydrocarbon molecules into smaller ones.

ANSWERS TO THE HOMEWORK PROBLEMS

Diagrams will vary but should demonstrate the student's grasp of how carbon bonds. Carbon atoms are always surrounded by *four* bonds whether the bonds are single, double, or triple bonds. Students should also show that isomers are true rearrangements of carbon atoms (i.e., branching configurations) rather than simple "twisted" versions of identical molecules. Two isomers of octane and octene are shown below.

octane isomers

octene isomers

ANSWERS TO THE END-OF-THE-WEEK REVIEW QUIZ

1. IVB
2. covalent
3. organic compounds
4. recycled
5. bacteria
6. coal
7. crude oil
8. refinery
9. distilled
10. plastics
11. crude oil
12. are not
13. true
14. oxygen
15. polymerization
16. D
17. C
18. B
19. E
20. A

CH11 FACT SHEET

CARBON COMPOUNDS AND PETROCHEMICALS

CLASSWORK AGENDA FOR THE WEEK

(1) Draw structural formulas and build models of carbon compounds.
(2) Discover the kinds of rock that serve as crude oil reservoirs and draw a diagram to show how petroleum is refined.
(3) Determine the number of calories in a piece of charcoal.
(4) Break up hydrocarbon molecules by catalytic cracking.

Carbon is a unique chemical element. Atoms of carbon (in Family IVB of *The Periodic Table of the Elements*) neither lose nor gain electrons to form **ionic bonds**. Carbon, like atoms of every member of its family, form **covalent bonds**. That is, carbon atoms "share" their outer shell electrons with the electrons of other atoms. Because carbon can share its electrons with the atoms of many other elements, the number of possible **carbon compounds** is enormous. Carbon is named after the Latin word for coal: *carbo*. It is found in cooking charcoal, the graphite tip of your pencil, and in diamonds. Carbon forms the "backbone" of all **hydrocarbon** molecules like **octane** which are used as fuels. Carbon also joins with oxygen, hydrogen, and nitrogen to form most of the important molecules that give life to every living organism on this planet. Carbon compounds that can be found in living organisms are called **organic compounds**.

When living things die their chemical elements are recycled through the environment. Plants grow. Animals eat plants. Animals die and decompose. Animal and plant remains are broken down by bacteria in the soil to nourish more growing plants. Sometimes the carbon compounds that make up dead plants and animals are preserved in the soil. Carbon compounds can be preserved for ages and can turn up almost anywhere. **Coal** and **crude oil** are the preserved remains of plants and animals that died millions of years ago. Crude oil is also called **petroleum** or **fossil fuel**. The atoms in a chunk of charcoal were once part of a prehistoric plant. The atoms in gasoline and plastic products were probably once present in the body of a dinosaur.

Petroleum is processed to make **petrochemicals**. You are surrounded by them! You are probably sitting on petrochemicals, writing with petrochemicals, and wearing combinations of them! **Plastics**, just one of the many products we get from petroleum, are made of petrochemicals.

As mentioned above, crude oil is the product of **organic compounds** that have been "cooking" in the earth for millions of years. They are the remains of living organisms that were buried deep beneath the surface of the earth. The primary chemical elements contained in organic compounds are carbon, hydrogen, oxygen, and nitrogen. Because it forms covalent bonds, carbon tends to form "chains" or "rings" of atoms that can be put together like Lego™ blocks. The unique ability of carbon atoms to form chains and rings allows **petrochemists** to create a host of amazing compounds such as **nylon**, **synthetic rubber**, and **polyester**. These substances are called polymers. A **polymer** is a long chain of carbon molecules called **monomers**. Figure 1 is an illustration of how several monomers bond to form a polymer. The chemical reaction used to connect the monomers is called a **polymerization reaction**.

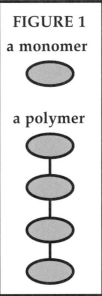

FIGURE 1

a monomer

a polymer

Crude oil found underground is pumped to the surface at oil wells then transported to a petroleum factory called a **refinery**. At a refinery, the oil is "distilled" to separate the organic compounds into **fractions**. Some of the organic compounds in crude oil are heavy while others are light. The different weights of these compounds allow chemists to isolate various petroleum products such as **wax, kerosene, asphalt, jet fuel, diesel fuel, heating oil**, and **gasoline** by a process of **distillation**.

Homework Directions

1. Write the structural formulas of octane and octene.
2. Draw three (3) isomers of octane.
3. Draw two (2) isomers of the octene.

Assignment due: _____

_____   _____   ____/____/____
Student's Signature   Parent's Signature   Date

CH11 Lesson #1

CARBON COMPOUNDS AND PETROCHEMICALS

Work Date: ____/____/____

LESSON OBJECTIVE

Students will draw structural formulas and build models of carbon compounds.

Classroom Activities

On Your Mark!

Display a piece of coal or charcoal, a bottle of mineral oil, and several plastic objects. Inform students that they will be studying the element that makes the existence of all of these substances possible: **carbon**.

Review the atomic structure of the carbon atom emphasizing the four outer shell electrons used to form covalent bonds with other atoms. Draw the Bohr electron shell diagram for carbon shown in Illustration A and the accompanying structural formulas showing the **single**, **double** and **triple bonded hydrocarbon chains** formed by carbon atoms. Explain that chains of carbon atoms with single bonds are called **saturated hydrocarbons** because these molecules hold a maximum number of hydrogen atoms (i.e., found in saturated fats). Double and triple bonded

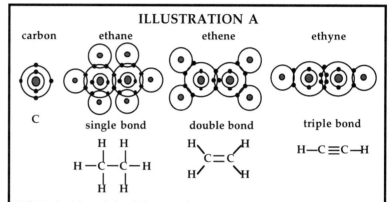

ILLUSTRATION A

carbon   ethane   ethene   ethyne

C   single bond   double bond   triple bond

NOTE: Build models of these molecules using molecular model kits (if available). Models will allow students to see the restrictions in molecular vibrational motion imposed by the double and triple bond structures. Explain that each "stick" bonding two atoms in the **structural formula** represents a pair of electrons. Note that each bonded carbon has 8 electrons in its outer shell although double and triple bonds restrict the motion of those electrons between "sharing" atoms.

hydrocarbons are called **unsaturated hydrocarbons** (i.e., in polyunsaturated oils). Give a brief lecture about the history of organic chemistry using the Teacher's Agenda and Content Notes. Have students copy your notes on Journal Sheet #1.

Get Set!

Explain that single bonded hydrocarbons belong to a family called the **alkane series**. **Alkenes** refer to carbon chains with double bonds. **Alkynes** refer to carbon chains with triple bonds.

Go!

Have students perform the modeling activity described in Figure A on Journal Sheet #1.

Materials

molecular model kits (if available) or toothpicks and several colors of clay

145

CH11 Journal Sheet #1

CARBON COMPOUNDS AND PETROCHEMICALS

FIGURE A

<u>Directions</u>: Use the materials given to you by your instructor to construct models of each of the hydrocarbon chains listed. **Alkanes** contain single bonds only. **Alkenes** need contain just one double bond. **Alkynes** need contain just one triple bond. Draw the structural formulas for each of the molecules you build.

propane C_3H_8	propene C_3H_6	propyne C_3H_4
butane C_4H_{10}	butene C_4H_8	butyne C_4H_6
pentane C_5H_{12}	pentene C_5H_{10}	pentyne C_5H_8
hexane C_6H_{14}	hexene C_6H_{12}	hexyne C_6H_{10}
heptane C_7H_{16}	heptene C_7H_{14}	heptyne C_7H_{12}

CH11 Lesson #2

CARBON COMPOUNDS AND PETROCHEMICALS

Work Date: ____/____/____

LESSON OBJECTIVE

Students will discover the kinds of rocks that serve as crude oil reservoirs and draw a diagram to show how petroleum is refined.

Classroom Activities

On Your Mark!

Before beginning a discussion of petroleum, review the structural diagram of the alkane butane. Draw Illustration B on the board and have students copy your diagram on either Journal Sheet #1 or #2. Explain that some carbon compounds have the same chemical formula but different structural formulas. Such compounds are called **isomers**. This feature of hydrocarbons is important because it is a molecule's structure that determines its physical and chemical properties. Note that both molecules in this illustration have the formula C_4H_{10}. However, butane has a different melting and boiling point than isobutane. Read over the Homework Assignment and be sure students understand the concept of an isomer.

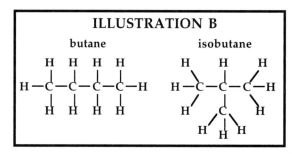

Get Set!

Explain that **petroleum** is the remains of living organisms that lived long ago. The petroleum industry explores the world for likely sites of petroleum deposits which are found both on land and on the continental shelf. Directly out of the ground, petroleum is "crude" or a jumble of many different hydrocarbons. At a **refinery** the mixture is "stored" then "separated" by **fractional distillation**. The fractions are further "treated" to rid them of contaminating molecules (i.e., sulphates) and "converted" into desired molecules that can be combined and recombined to make thousands of different products from gasoline to plastics. Draw Illustration C on the board and explain how distillation is used to separate the components of crude oil. Have students copy your diagram on Journal Sheet #2.

ILLUSTRATION C

low temperature — gas

aviation fuel, gasoline

turbo jet fuel, kerosene

diesel fuel

lubricating oils

high temperature crude oil — waxes

Light fractions of the crude oil mixture vaporize, rise through the **fractionating column**, and are siphoned off for use.

Go!

Have students perform the brief activity described in Figure B on Journal Sheet #2. Students will discover that the more porous rocks (i.e., sandstone and limestone) hold the most oil. Petroleum reservoirs are found in this type of rock **sedimentary rock**.

Materials

rock samples (sandstone, limestone or chalk, granite, etc.), mineral or baby oil, petri dishes, medicine droppers, digital or stopwatch

147

CH11 JOURNAL SHEET #2
CARBON COMPOUNDS AND PETROCHEMICALS

FIGURE B

Directions: (1) Place each rock sample in a separate beaker. (2) Use the medicine dropper to place one drop of oil on each sample and record the time it takes for the entire drop to be absorbed by the rock. (3) Stop timing after five minutes. (4) Which rock absorbed the oil the fastest? Explain why this type of rock is more likely to be a good reservoir for crude oil.

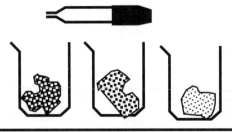

CARBON COMPOUNDS AND PETROCHEMICALS

Work Date: ____/____/____

LESSON OBJECTIVE

Students will determine the number of calories in a piece of charcoal.

Classroom Activities

On Your Mark!

The considerable amount of energy stored in fossil fuels can be demonstrated in a number of ways. The demonstration described in Illustration D <u>must be rehearsed prior to class and done outdoors in order to insure the safety of the students and instructor</u>. Should you choose to forego this demonstration simply begin discussion with the history of coal using the information provided in the Teacher's Agenda and Content Notes.

Get Set!

Review the concept of the **calorie** discussed in unit CH4—*Heat and Energy Transfer*. Ask students if they recall the amount of energy released by the burning soda cracker. Point out that the soda cracker was also made of carbon and hydrogen just like any hydrocarbon. Remind students that the soda cracker came from a plant much like the plants that existed on earth millions of years ago: plants that were buried and fossilized into the coal used to make charcoal.

ILLUSTRATION D

<u>Directions</u>: (1) Construct a "cannon" from a wooden dowel by drilling a hole lengthwise through the stock. Drill a thinner second hole through the side into the barrel. (2) Place the cannon on the ground with the open end elevated by a brick or block of wood. (3) Cap the mouth of the cannon LIGHTLY with a cork. (4) Introduce a SINGLE DROP of gasoline into the barrel from the side hole as shown and cap it with your thumb allowing the gas to vaporize. (5) Remove your thumb and light a match to the side opening.

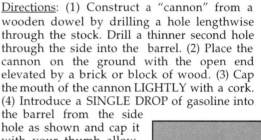

GENERAL SAFETY PRECAUTIONS

Perform this demonstration outdoors. Have students stand a safe distance away (i.e., 20 feet) from the cannon. Wear goggles.

Go!

Have students perform the demonstration described in Figure C on Journal Sheet #3. If available, have them compare their results to those obtained in the activity performed in Lesson #4 of unit CH4—*Heat and Energy Transfer*.

Materials

ring stands, ring clamps, thermometers, metal spoons, soda cans, water, charcoal, hammer (to prepare small pieces), supplies mentioned in Illustration D (if desired)

CH11 JOURNAL SHEET #3

CARBON COMPOUNDS AND PETROCHEMICALS

FIGURE C

<u>Directions</u>: (1) Pour 100 ml of water from a beaker into a soda can. (2) Secure the soda can with two ring clamps as shown. (3) Lower a thermometer into the can just below the surface of the water. (4) Dip a small piece of charcoal (about one cubic centimeter in size) in lighter fluid and place it in a spoon secured to the ring stand under the can. (5) Record the temperature of the thermometer. (6) Light a match to the soaked charcoal and allow it to burn on its own. (7) Record the temperature reading when the charcoal is completely burned. (8) Since one calorie is the amount of energy needed to raise the temperature of water one degree Celsius, the number of calories in the charcoal is equal to the rise in temperature caused by the burning x100 (i.e., the volume of water in the can).

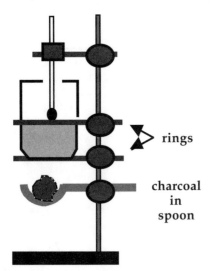

rings

charcoal
in
spoon

GENERAL SAFETY PRECAUTIONS
Wear goggles to protect your skin and eyes when working with a flame. Do not inhale the vapors of the burning charcoal. Do not touch any part of the equipment without heat-resistant gloves or tongs. Clean up only after the apparatus is cool.

CH11 Lesson #4

CARBON COMPOUNDS AND PETROCHEMICALS

Work Date: _____/_____/_____

LESSON OBJECTIVE

Students will break up hydrocarbon molecules by catalytic cracking.

Classroom Activities

On Your Mark!

Draw long-chain molecules of the alkane series like those shown in Illustration E and have students copy them on Journal Sheet #4. Reiterate that crude oil is a "soupy mixture" of organic compounds containing long and short chain hydrocarbons and isomers of every kind. Following the separation of the molecules into fractions by fractional distillation, **organic chemists** can also convert large molecules into smaller ones by **catalytic cracking**. That is, they use heat and a metal **catalyst** to break apart long molecules. High octane gasoline is produced in this way.

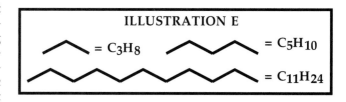

ILLUSTRATION E

$= C_3H_8$ $= C_5H_{10}$

$= C_{11}H_{24}$

Get Set!

Point out that mineral oil is a product of petroleum and that it contains long-chained hydrocarbons. In the first part of the experiment described in Figure D on Journal Sheet #4 students will vaporize and condense the mineral oil, capturing it by water displacement. In the second part of the experiment they will show how a metal catalyst (steel wool) speeds up the "cracking" of these large molecules. If available, you can use a dilute potassium permanganate solution to indicate the presence of the butylene molecules produced by this catalytic cracking procedure. This can be done by allowing some of the water in the beaker to mix with the captured contents following each phase of the experiment. Add a drop of dilute potassium permanganate to each solution. The first solution will not decolorize the potassium permanganate. However, potassium permanganate does decolorize in the presence of the unsaturated hydrocarbons (e.g., butylene) mixed in with the solution produced in the second phase of the experiment. As students complete each phase of the experiment you can go around the room performing this simple demonstration.

Go!

Review the General Safety Precautions described in Figure D on Journal Sheet #4 and give students ample time to complete the experiment.

Materials

Bunsen burners, 500 ml beakers, large test tubes, rubber stopper–rubber tubing assembly, water pans, ring stand and clamps, mineral oil, steel wool, dilute potassium permanganate (if available)

CH11 JOURNAL SHEET #4

CARBON COMPOUNDS AND PETROCHEMICALS

FIGURE D

<u>Directions</u>: (1) Pour 5 milliliters of mineral oil into a large test tube. (2) Stopper and secure the test tube at an angle as shown BUT WITHOUT THE STEEL WOOL. In the first part of this experiment you will heat the mineral oil without the steel wool. (3) Fill a pan with water and submerge a 500 ml beaker in the pan to fill it. Invert the beaker and insert the rubber tubing from the stopper into the beaker. (4) Light the Bunsen burner under the mineral oil and SLOWLY HEAT the oil over a SMALL FLAME until oil droplets collect inside the beaker. Did the oil undergo a chemical change or was it simply vaporized and recondensed? Chemical or physical change? (5) Give the apparatus a few moments to cool then uncork the large test tube and insert a small wad of steel wool. (6) Rinse and dry the beaker and repeat Step #3. (7) Light the Bunsen burner under the steel wool and preheat it for one minute. (8) Light the Bunsen burner under the mineral oil and observe the gas produced this time. Was the same amount of gas produced? Do as many mineral oil droplets appear or is there more clear gas? Was this a physical or chemical change?

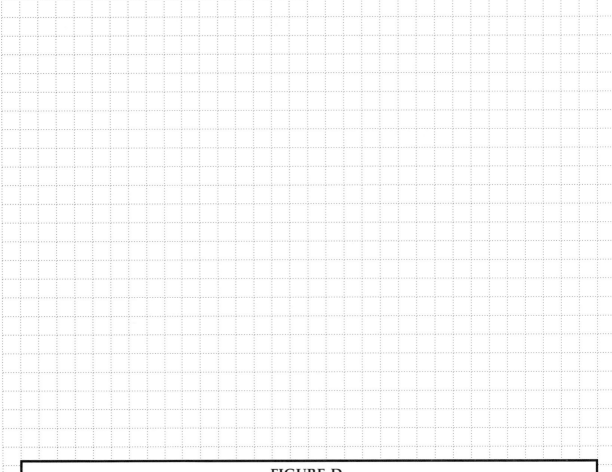

steel wool

mineral oil

GENERAL SAFETY PRECAUTIONS

Wear safety goggles. Do not touch any part of the equipment while it is hot. Use heat-resistant gloves or tongs. Do not inhale the gases produced in this experiment.

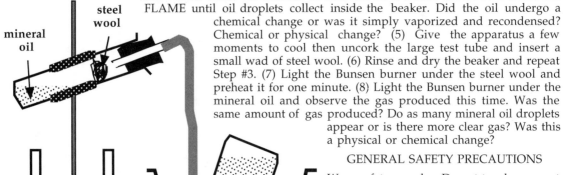

CH11 Review Quiz

Directions: Keep your eyes on your own work.
Read all directions and questions carefully.
THINK BEFORE YOU ANSWER!
Watch your spelling, be neat, and do the best you can.

CLASSWORK   (~40): _____
HOMEWORK   (~20): _____
CURRENT EVENT (~10): _____
TEST   (~30): _____

TOTAL   (~100): _____
(A ≥ 90, B ≥ 80, C ≥ 70, D ≥ 60, F < 60)

LETTER GRADE: _____

TEACHER'S COMMENTS: _____

_____

CARBON COMPOUNDS AND PETROCHEMICALS

PARAGRAPH FILL-IN: Fill in the blanks using the vocabulary that appeared on your Fact Sheet.
10 points

Carbon is a unique chemical element. Carbon is in Family (1)_____ of *The Periodic Table of the Elements*. Carbon neither loses nor gains electrons to form ionic bonds but forms (2)_____ bonds instead. Because carbon can "share" its electrons with almost any other type of atom, the number of possible carbon compounds is enormous. Carbon compounds found in living things are called (3)_____ _____.

When organisms die their chemical elements are (4)_____ through the environment. Plants grow, animals eat plants, animals die and decompose and their compounds are broken down by (5)_____ in the soil. Recycled elements go on to nourish more growing plants. Sometimes, the carbon compounds that make up dead plants and animals are preserved in the soil. (6)_____ and (7)_____ _____ are the remains of animals and plants that died millions of years ago. The atoms in a piece of charcoal were once part of a prehistoric plant. The atoms in gasoline and plastic were once present in the body of a dinosaur.

Crude oil is transported to a petroleum factory called a (8)_____. There, the oil is (9)_____ to separate the organic compounds into fractions. Some organic compounds are heavy while others are light. This allows petrochemists to isolate different kinds of substances to make many products including waxes, kerosene, asphalt, jet fuel, diesel fuel, and heating oil. (10) _____ are the most commonly used solid petroleum products.

CH11 Review Quiz (cont'd)

TRUE–FALSE FILL-IN: If the statement is true, write the word TRUE. If the statement is false, change the underlined word to make the statement true. *10 points*

_____11. Plastics are made of chemicals that come from petroleum which is a product of <u>gasoline</u>.

_____12. Plastics <u>are</u> the only substances we get from petroleum.

_____13. Hydrocarbons <u>are</u> the remains of living organisms that have been buried deep beneath the surface of the earth for a long time.

_____14. The main chemical elements contained in organic compounds are carbon, hydrogen and <u>helium</u>.

_____15. Nylon, dacron, and polyester are the products of a(n) <u>acid-base</u> reaction.

MATCHING: Choose the letter of the word or phrase on the right that best describes the refinery process on the left. *10 points*

_____ 16. exploration (A) maintain a supply

_____ 17. separation (B) make smaller molecules from bigger ones

_____ 18. conversion (C) distill

_____ 19. treatment (D) search and find

_____ 20. storage (E) purify

CH12

THE MOLECULES OF LIFE

TEACHER'S CLASSWORK AGENDA AND CONTENT NOTES

Classwork Agenda for the Week

1. Students will examine the adhesive and cohesive properties of water.
2. Students will draw and construct models of carbohydrate and fat molecules.
3. Students will draw and construct a model of a protein molecule.
4. Students will draw and construct a model of a DNA molecule.

Content Notes for Lecture and Discussion

What is life? This primary question has vexed biologists since ancient times and to this day scientists argue about the nature and "qualifications" of living things. Should we one day have computers able to do everything a living organism can do, will it be necessary to classify them as living things?

The word "life" was used by the Ancient Greeks to classify objects as diverse as plants, insects, worms, fish, birds, and man. The Greeks observed that all of these "living things" went through a "life cycle" from birth to death and unlike inanimate objects were able to reproduce their own kind. The invention of the microscope by the Dutch pioneer **Anton von Leeuwenhoek** (b. 1632; d. 1723) allowed the German botanist **Matthias Jakob Schleiden** (b. 1804; d. 1881) and physiologist **Theodor Schwann** (b. 1810; d. 1882) to discover that all living things were composed of cells. **Charles Robert Darwin** (b. 1809; d. 1882) gave biology its first compelling theory by explaining the concept of historical and biological evolution with his **theory of evolution by means of natural selection**. While Darwin did not understand the "mechanism" by which organisms transmitted their physical characteristics to their offspring, he predicted that such a mechanism would one day be found.

The progress made in the 18th and 19th centuries by physical and chemical scientists led students of biology to ask more sophisticated and pointed questions. Atmospheric gases were identified and isolated during this time. The production of oxygen by plants during **photosynthesis**—and the reverse process used by animals during **respiration**—served to support the theory of **conservation of matter and energy** proposed by **Anton Laurent Lavoisier** (b. 1743; d. 1794) and **Pierre Laplace** (b. 1749; d. 1827). They declared that biological systems obeyed physical law. However, the idea of a photosynthesis/respiratory cycle was not proposed until later. In 1828, the German chemist **Friedrich Wöhler** (b. 1800; d. 1882) synthesized urea from inorganic chemicals proving that a "vital force" was not necessary to create "living molecules" from "nonliving molecules." But it was the formulation of the principles of catalysis by the Swedish chemist **Jöns Jakob Berzelius** (b. 1779; d. 1848) that led to the birth of **organic** or **biochemistry**. Berzelius realized that chemicals present in living things, such as saliva and gastric juices, sped the chemical digestion of life-sustaining nutrients. Since then, the scope of biochemistry has widened to include all of the chemical reactions that take place in living things. The German physician **Ernst Felix Hoppe-Seyler** (b. 1825; d. 1895) founded and published the first journal of biochemistry—*Zeitschrift für Physiologische Chemie*—and made major contributions to the study of chlorophyll and hemoglobin: molecules without which life as we know it would be impossible. The photosynthesis/respiratory cycle was suggested in the work of the German chemist **Julius von Sachs** (b. 1832; d. 1897) and showed that **starch** present in plant cells was the product of a chemical reaction that involved the absorption of carbon dioxide from the atmosphere. This work led to the discoveries of the German botanist **Andreas Franz Wilhelm Schimper** (b. 1856; d. 1901) who demonstrated that plants stored energy radiated by the sun and that animals used that energy to sustain metabolic activities.

The ability to derive molecular structure from the diffraction patterns created by the bombardment of organic molecules with X-rays (i.e., X-ray crytallography) led to the elucidation of the structures of both **proteins** and **nucleic acids** during the 1950s. The discovery of the structure of proteins is credited to the American theoretical chemist **Linus Carl Pauling** (b. 1901; d. 1994). The American biologist **James Dewey Watson** (b. 1928) and the English microbiologist **Francis Harry Compton Crick** (b. 1916) shared the Nobel Prize in 1962 for their discovery of the structure of the DNA molecule. In 1953, Watson and Crick, while working at Cambridge University, constructed a model of the DNA molecule that allowed them to explain how hereditary information was encoded in the gene.

In Lesson #1, students will review the chemical equations for photosynthesis and respiration and examine the adhesive and cohesive properties of water—the universal solvent.

In Lesson #2, students will construct models of carbohydrate and fat molecules.

In Lesson #3, students will construct a model of a protein molecule.

In Lesson #4, students will draw and construct a model of a DNA molecule.

ANSWERS TO THE HOMEWORK PROBLEMS

Essays will vary but should demonstrate the student's understanding of the similarities between polymerization and dehydration synthesis. Both chemical reactions result in the synthesis of large molecules from smaller chemical units.

ANSWERS TO THE END-OF-THE-WEEK REVIEW QUIZ

1. C	6. B	11. B	ESSAY: Essays will vary but should emphasize
2. B	7. D	12. D	that both chemical reactions result in the
3. E	8. A	13. A	synthesis of large molecules from
4. A	9. C		smaller chemical units.
5. D	10. C		

CH12 Fact Sheet

THE MOLECULES OF LIFE

CLASSWORK AGENDA FOR THE WEEK

(1) Examine the adhesive and cohesive properties of water.
(2) Draw and construct models of carbohydrate and fat molecules.
(3) Draw and construct a model of a protein molecule.
(4) Draw and construct a model of a DNA molecule.

Four billion years ago, planet earth was not as it is appears today. The planet's crust and atmosphere were raging hot. Erupting volcanoes spotted the landscape, flooding the terrain with lakes of molten metal and filling the skies with tremendous amounts of **carbon dioxide** (CO_2), **ammonia** (NH_3), and **methane** (CH_4). There was little free **oxygen** (O_2) in the atmosphere. Most of that element was combined with hydrogen to form **water** (H_2O) or joined with metals to form metal oxides. About three billion years ago, however, things changed. The electrical energy from lightning discharges in the atmosphere helped to combine these basic building-block molecules into the raw materials needed to form living organisms. In the 1950s, American chemists **Stanley Lloyd Miller** (b. 1930) and **Harold Clayton Urey** (b. 1893; d. 1981) performed a series of experiments that showed how this "organic synthesis" could have occurred. Using a device similar to the one illustrated in Figure 1, Miller and Urey demonstrated that the large molecules that make up all living things can be made from carbon dioxide, ammonia, methane, and water. Large molecules like carbohydrates, fats, proteins, and nucleic acids can all be made from the substances that existed on the very primitive earth.

The science concerned with the chemistry of living organisms is called **biochemistry**.

Carbohydrates give living organisms energy to burn. They are made of simple sugars called **saccharides**. The simplest saccharide is **glucose** ($C_6H_{12}O_6$). Saccharides combine in a type of polymerization reaction called **dehydration synthesis**. During dehydration synthesis a molecule of water (2 hydrogen atoms and 1 oxygen atom) is removed from each glucose molecule to form a **polysaccharide** chain. A polysaccharide is a long chain of glucose molecules.

Fats—also called **lipids**—help to form an organism's protective tissues. Fats also serve as a secondary source of energy. Like carbohydrates, fats are also produced by dehydration synthesis. Hydrocarbon chains called "fatty acids" and a molecule of glycerol

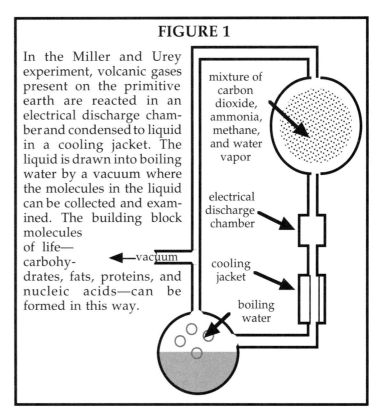

FIGURE 1

In the Miller and Urey experiment, volcanic gases present on the primitive earth are reacted in an electrical discharge chamber and condensed to liquid in a cooling jacket. The liquid is drawn into boiling water by a vacuum where the molecules in the liquid can be collected and examined. The building block molecules of life—carbohydrates, fats, proteins, and nucleic acids—can be formed in this way.

mixture of carbon dioxide, ammonia, methane, and water vapor

electrical discharge chamber

vacuum

cooling jacket

boiling water

($C_3H_8O_3$) are combined to make the substance we call fat. Water is a secondary product of this reaction just as it is in the production of carbohydrates.

Proteins give an organism its structure. Bones, muscles, and blood vessels are all made of proteins. The building blocks of proteins are molecules called **amino acids**. There are about twenty different amino acids in nature. Different combinations of amino acids give rise to the millions of proteins that exist in nature. Proteins are also produced by dehydration synthesis.

Nucleic acids carry the "hereditary features" of every living organism. The instructions for assembling an organism's particular physical characteristics (i.e., eye color, hair color, body shape) are passed from one generation to the next by these long-chained organic compounds. Nucleic acids are formed by dehydration synthesis from smaller units called **nucleotides**. Nucleotides are arranged in a coded sequence and held together in a chain made of **phosphates** and **sugars**. The complex **macromolecule** that carries the hereditary features of every living plant or animal on our planet is called **deoxyribonucleic acid**: **DNA**.

Homework Directions

In an essay of about 100 words explain why a knowledge of polymerization reactions is important to the study of biochemistry.

Assignment due: _____

_____   _____   ____/____/____
        Student's Signature   Parent's Signature   Date

THE MOLECULES OF LIFE

Work Date: ____/____/____

LESSON OBJECTIVE

Students will examine the adhesive and cohesive properties of water.

Classroom Activities

On Your Mark!

Begin discussion by reminding students that the sun is the primary source of energy that sustains life on our planet. Review the chemical equation for **photosynthesis** and draw the diagram in Illustration A on the board. Have students copy the equation and diagram on Journal Sheet #1. Point out that **water** is essential to the survival of all living things. Explain that water is the **universal solvent** in which most **biochemical reactions** take place.

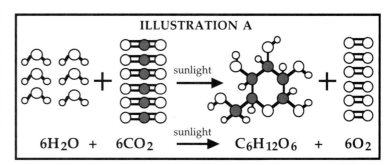

ILLUSTRATION A

$$6H_2O + 6CO_2 \xrightarrow{\text{sunlight}} C_6H_{12}O_6 + 6O_2$$

Get Set!

Splash some water droplets on a glass window and point out that water sticks to a variety of different substances. **Adhesion** is the term used to refer to the force that causes different substances to stick to one another. Dangle a drop of water from your finger tip and point out that water molecules form a spherical droplet indicating that water molecules can also stick to each other. **Cohesion** is the term used to refer to the force that causes molecules of a particular substance to stick to molecules of the same substance. Water being absorbed by the root hairs of a plant is the result of adhesion. **Surface tension** created at the surface of a cup of water or around a water droplet suspended from an icicle is an example of cohesion. Draw Illustration B and explain that water is a **bipolar molecule**. The oxygen in water tends to have negative electrons flying around it while the hydrogens remain relatively positive. The opposite "poles" of water molecules are attracted to one another. The idea that molecules could be attracted to one

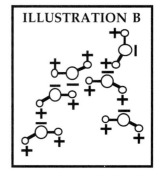

ILLUSTRATION B

another in this manner was first proposed by the Dutch physicist **Johannes Diderik van der Waals** (b. 1837; d. 1923) in 1873. The forces between the molecules are called **van der Waals' forces**.

Go!

Give students sufficient time to have fun with the demonstrations described in Figure A and Figure B.

Materials

water, paper cups or beakers, string, paper clips, thumb tacks.

CH12 JOURNAL SHEET #1

THE MOLECULES OF LIFE

FIGURE A

Directions: (1) Place paper towels down to ease clean up. (2) Fill a paper cup with water. (3) Soak a length of string at least 3 feet long in the cup. (4) Have a classmate hold a second paper cup down on the paper towel. (5) Insert the soaked string into the two cups, securing the string under a finger. (6) Slowly

pour the water from one cup into the other allowing the liquid to run down the taut length of string. (7) How does this demonstrate the cohesive properties of water?

FIGURE B

Directions: (1) Fill a paper cup with water so that the water "bulges" over the rim of the cup as shown. (2) Place a paper clip or tack at the edge of the rim and try to push it on to the surface without breaking the **surface tension** of the water. (3) "Float" as many clips or tacks as you can.

CH12 Lesson #2

THE MOLECULES OF LIFE

Work Date: ___/___/___

LESSON OBJECTIVE

Students will draw and construct models of carbohydrate and fat molecules.

Classroom Activities

On Your Mark!

Review the properties of water and point out that water is a common reactant or product in many biochemical reactions. While water acts as the universal solvent for many biochemical reactions it is also the product of **dehydration synthesis**. During dehydration synthesis a water molecule is lost as two molecules are linked together in a chain. Carbohydrates, fats, proteins, and nucleic acids are all formed by dehydration synthesis.

Get Set!

Use Illustration C and ball-and-stick or clay-and-toothpick models to show how dehydration synthesis occurs to form a carbohydrate molecule. Use Illustration D and ball-and-stick or clay-and-toothpicks to show how dehydration synthesis occurs to form a fat molecule.

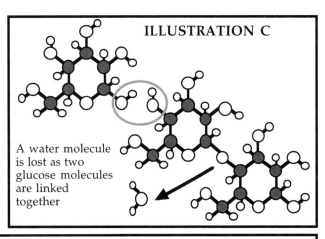

ILLUSTRATION C

A water molecule is lost as two glucose molecules are linked together

Go!

Give students ample time to draw accurate diagrams and build accurate models of carbohydrates and fats formed by dehydration synthesis. Explain that fats are used to make soap because the cell membranes of skin cells are made of lipid; and since lipids cohere, fats can be used to wash dead skin cells

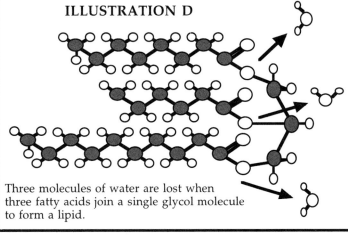

ILLUSTRATION D

Three molecules of water are lost when three fatty acids join a single glycol molecule to form a lipid.

from the body. Allow students to have "Fun with Fats" as directed in the activity on Journal Sheet #2.

Materials

ball-and-stick or clay-and-toothpick models to make molecules, liquid dish soap and glycerol, ring clamps, water pans

CH12 Journal Sheet #2

THE MOLECULES OF LIFE

Draw several more glucose molecules like the one shown and diagram how they are linked into a chain by dehydration synthesis.

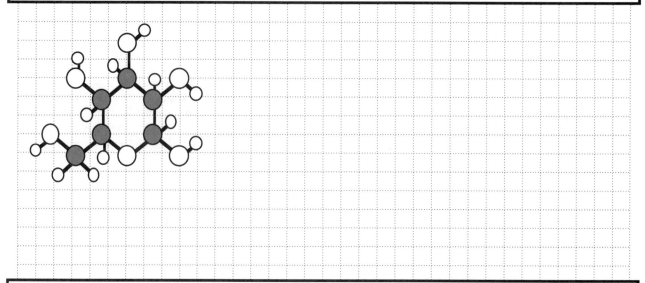

Draw several more fatty acid and glycol molecules like those shown and diagram how they are linked into a chain by dehydration synthesis.

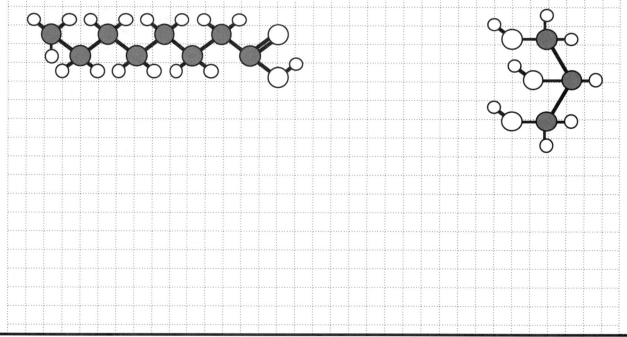

FUN WITH FATS

<u>Directions</u>: (1) Add several milliliters of glycerol to a pie pan or shallow water dish filled 2 centimeters deep with liquid dish soap. (2) Use a ring clamp to make soap bubbles that you can examine. (3) Find a way to stick your finger through the bubbles without bursting them. Why do you think fats work well as the membranes that keep living cells together?

THE MOLECULES OF LIFE

Work Date: ____/____/____

LESSON OBJECTIVE

Students will draw and construct a model of a protein molecule.

Classroom Activities

On Your Mark!

Explain the significance of **proteins** in forming the basic structures and catalytic enzymes that make life possible. Point out that proteins are chains of **amino acids** like **glycine, alanine,** and **valine** that link together to form the vast variety of proteins that exist in nature.

Get Set!

Draw Illustration E on the board to show how proteins are formed by dehydration synthesis. Explain that the bond formed between the terminal hydroxyl (–OH) and amino (–NH$_2$) groups are called **peptide bonds**. Proteins are **polypeptides**.

Go!

Give students ample time to draw and construct a model of a protein. Have them alter the sequence of the three amino acids to show how the polypeptide molecule twists into different shapes depending upon the amino acid sequence.

ILLUSTRATION E

A molecule of water is lost in the formation of each peptide bond. The shape of a particular protein is determined by the sequence of amino acids joined together. Since there are about 20 amino acids in nature that can combine in any sequence the possible number of proteins that could exist is virtually limitless.

Materials

ball-and-stick or clay-and-toothpick models to make molecules

CH12 JOURNAL SHEET #3

THE MOLECULES OF LIFE

Draw a diagram to show how the amino acid molecules glycine, alanine, and valine will link together to form a protein by dehydration synthesis.

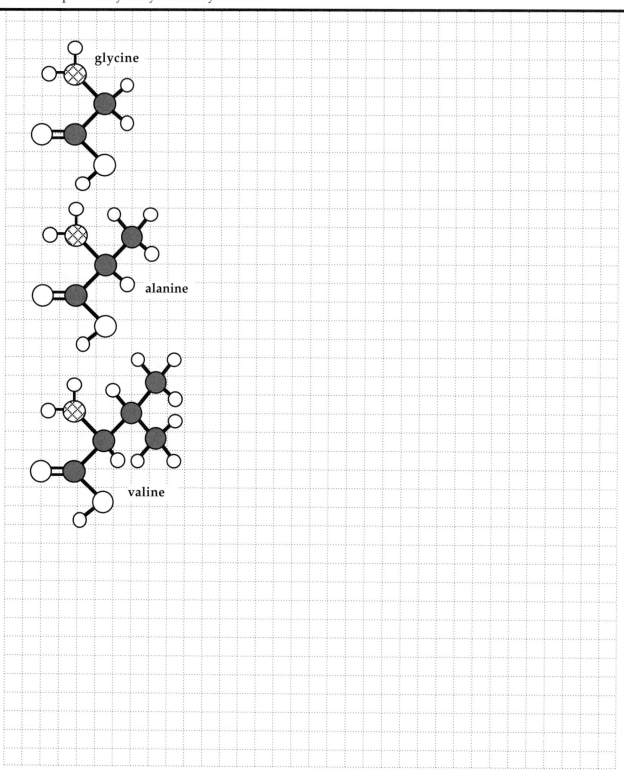

glycine

alanine

valine

THE MOLECULES OF LIFE

Work Date: ____/____/____

LESSON OBJECTIVE

Students will draw and construct a model of a DNA molecule.

Classroom Activities

On Your Mark!

Begin the lesson with a discussion of the similarities between the organic molecules studied in Lessons #1, #2, and #3. Ask: How are carbohydrates, fats, and proteins similar? Answer: They are all present in living things and they contain the elements carbon, oxygen, and hydrogen. Have students discuss how these molecules are different. In addition to contrasting the shapes of the molecules, some will point out that proteins contain nitrogen in addition to carbon, oxygen, and hydrogen. Explain that nucleic acids—other molecules of life—contain the element phosphorus. Point out that nucleic acids, like other organic molecules, are formed by dehydration synthesis.

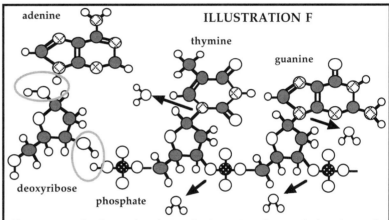

ILLUSTRATION F

adenine

thymine

guanine

deoxyribose

phosphate

Two water molecules are lost during the formation of a strand of nucleic acid as a phosphate and nucleotide are joined to a deoxyribose molecule. The deoxyribose-phosphate pair serves as the "backbone" of the DNA molecule. The sequence of nucleotides along the strand determines the protein to be made. A sequence of 3 nucleotides "codes" for a specific amino acid to be linked into a particular protein chain.

Get Set!

Remind students that proteins are responsible for giving an organism its structure and the catalytic enzymes that allow it to carry on important biochemical reactions. Ask students to explain how a living organism passes on the instructions for manufacturing those proteins to its offspring. Discuss the terms gene and heredity. A **gene** is a unit of inherited material. There are genes for eye color, hair texture, and sugar metabolism. **Heredity** is the transmission of traits from parent to offspring. Explain that a miraculous macromolecule, **deoxyribonucleic acid** or **DNA**, carries this "code of life." Draw Illustration F to show how dehydration synthesis results in the formation of a single strand of DNA. Point out that two "matching strands" of DNA—joined by van der Waals' forces—can be found in living cells in structures called **chromosomes**.

Go!

Give students ample time to draw and construct a model of DNA. They can do this on a large sheet of butcher paper, each student taking a section and connecting their section to those of other students.

Materials

ball-and-stick or clay-and-toothpick models to make molecules, butcher paper

CH12 JOURNAL SHEET #4

THE MOLECULES OF LIFE

Follow your instructor's directions so that you can link together these molecules to form a nucleic acid chain.

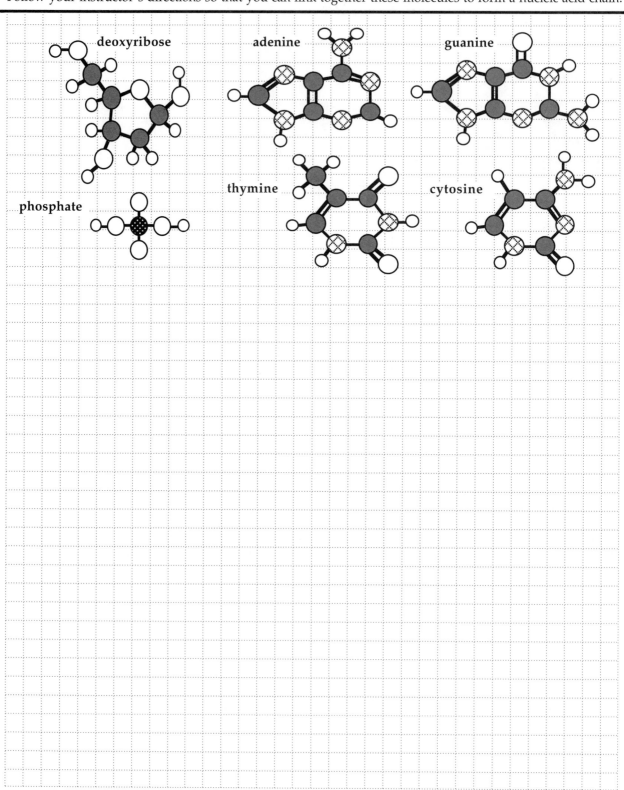

deoxyribose

adenine

guanine

phosphate

thymine

cytosine

CH12 REVIEW QUIZ

Directions: Keep your eyes on your own work.
Read all directions and questions carefully.
THINK BEFORE YOU ANSWER!
Watch your spelling, be neat, and do the best you can.

TEACHER'S COMMENTS: _____
_____

THE MOLECULES OF LIFE

MULTIPLE CHOICE: Choose the letter of the word or phrase that best completes the sentence or answers the question. *10 points*

_____ 1. Which of the following was probably NOT present in earth's primitive atmosphere?
(A) carbon dioxide
(B) methane
(C) polyvinylchloride
(D) water
(E) ammonia

_____ 2. Which two scientists discovered that large organic molecules can be made from simpler molecules that were probably present in earth's primitive atmosphere?
(A) Watson and Crick
(B) Miller and Urey
(C) Orville and Wilbur Wright
(D) Michelson and Morley
(E) Kirchoff and Bunsen

_____ 3. Which of the following are large molecules that make up living organisms?
(A) carbohydrates
(B) fats
(C) proteins
(D) nucleic acids
(E) all of the above

_____ 4. The science concerned with the chemistry of living organisms is called _____?
(A) biochemistry
(B) geochemistry
(C) physical chemistry
(D) neurochemistry
(E) astrophysics

_____ 5. When did the "organic synthesis" of life's basic building-block molecules probably first begin on planet earth?
(A) about 3 hundred years ago
(B) about 3 thousand years ago
(C) about 3 million years ago
(D) about 3 billion years ago
(E) about 3 trillion years ago

MATCHING: Choose the letter of the molecule appearing on the right that is a "building block" for each larger molecule listed on the left. *8 points*

_____	6.	carbohydrate	(A) amino acid
_____	7.	fats	(B) saccharides
_____	8.	proteins	(C) nucleotides
_____	9.	nucleic acids	(D) glycerol

MATCHING: Choose the letter of the term or phrase appearing on the right that best describes the primary use of each larger molecule listed on the left. *8 points*

_____	10.	carbohydrate	(A) hereditary material
_____	11.	fats	(B) insulation and protection
_____	12.	proteins	(C) source of fuel
_____	13.	nucleic acids	(D) shape and structure

ESSAY: In one or two brief sentences explain why dehydration synthesis is a type of polymerization reaction. *4 points*

_____

_____

_____

_____

_____

NUTRITION CHEMISTRY

TEACHER'S CLASSWORK AGENDA AND CONTENT NOTES

Classwork Agenda for the Week

1. Students will keep a daily record of the foods they eat and analyze their nutritional value.
2. Students will test a food sample for the presence of protein.
3. Students will test a food sample for the presence of starch.
4. Students will convert starch to simple sugar.

Content Notes for Lecture and Discussion

The dietary theories of the great Greek authorities **Hippocrates** (b. 460 B.C.; d. 377 B.C.) and **Galen** (b. 129; d. 200) were the first to guide physicians in the use of food to conserve health and fight disease. Hippocrates taught that all habits including eating and drinking habits were to be exercised with moderation, that one should expose one's self to the freshness of the air, and that nature would take its course in balancing the body "humours." Galen adopted these ideas and for the next thousand years food was classified according to its physical qualities such as temperature and moistness. This was consistent with the notion that the health of the body was determined by the balanced flow of the body's humoral fluids which Galen described as "phlegm, blood, choler, and yellow bile." These terms were used more to describe the states of illness resulting from an imbalance of the humors than to actual substances circulating throughout the body. Galen's ideas were challenged by the work of the Belgian physician **Andreas Vesalius** (b. 1514; d. 1564) whose in-depth anatomical studies published in 1543 contradicted Galen's grasp of anatomy. In 1628, the English physician **William Harvey** (b. 1578; d. 1657) demonstrated how the heart and blood vessels circulated blood throughout the body which allowed physicians to discount the humoral hypothesis. In addition, the advancement of quantitative chemistry led 19th century scientists to view dietary laws in light of food substances involved in the body's **metabolism**.

The German chemist **Justus von Liebig** (b. 1803; d. 1873) was among the first to examine excreted products like urea and carbon dioxide in relation to ingested foodstuffs like animal protein and plant sugar. The French physiologist **Claude Bernard** (b. 1813; d. 1878) demonstrated the digestive functions of the pancreas and showed how the liver broke down foods in the synthesis of glycogen. Bernard is considered the father of experimental medicine. The importance of amino acids as the building blocks of proteins was demonstrated by the American nutritionist and biochemist **Elmer Verner McCollum** (b. 1879; d. 1967). McCollum also elucidated the role of fats, vitamins, and minerals present in the diet of animals.

In Lesson #1, students will devise a table that will allow them to keep a daily record of the foods they eat. During the course of the unit, they will analyze the nutritional value of the foods they ingest.

In Lesson #2, students will test a food sample for the presence of protein.

In Lesson #3, students will test a food sample for the presence of starch.

In Lesson #4, students will convert starch to simple sugar.

CH13 Content Notes *(cont'd)*

ANSWERS TO THE HOMEWORK PROBLEMS

Students' tables will be as varied as their diets. However, they should write an honest essay about their personal eating habits and reflect upon the changes they need to make to develop healthier eating habits. Check to see that they have accurately identified the proper nutrients in the most common foodstuffs (i.e., breakfast cereal is a major source of carbohydrates, not fats and proteins).

ANSWERS TO THE END-OF-THE-WEEK REVIEW QUIZ

Answers to questions #1, #2, and #3 may be in any order.

1. make new products	6. carbohydrates	11. fats	16. true	21. C
2. repair old parts	7. true	12. proteins	17. minerals	22. E
3. produce energy	8. true	13. complete	18. true	23. D
4. true	9. water	14. incomplete	19. A	24. B
5. true	10. 1,000	15. true	20. F	

CH13 FACT SHEET

NUTRITION CHEMISTRY

CLASSWORK AGENDA FOR THE WEEK

(1)   Keep a daily record of the foods you eat and analyze their nutritional value.
(2)   Test a food sample for the presence of protein.
(3)   Test a food sample for the presence of starch.
(4)   Convert starch to simple sugar.

Living organisms, including humans, need a balanced diet in order to live a healthy life. The Greek physician **Hippocrates** (b. 460 B.C.; d. 377 B.C.) was well aware of the need to eat healthy foods to prevent ill health and fight off disease. In fact the word "diet" comes from the Greek word *diaita* which means "way of life." Food contains substances the body needs to carry on **metabolic activity**. **Metabolism** refers to the chemical processes that result in the manufacture of new products, the repairing of old parts, and the release of energy. The building of new body parts is called **anabolism**. The breaking down of substances in the body—as in the process of digestion— is called **catabolism**. **Nutrients** are the usable substances in food that help living organisms to accomplish these important functions. The six nutrients required by all living things are **carbohydrates, fats, proteins, vitamins, minerals,** and **water**.

Carbohydrates are an organism's primary source of energy. Carbohydrates are found in fruits, vegetables, and grain products. **Sugars** and **starches** are the two main types of carbohydrates. Sugars are simple molecules made of carbon, hydrogen, and oxygen. Plants produce sugar during **photosynthesis** by trapping the radiant energy of the sun inside the chemical bonds that hold together the atoms of the sugar molecule. Starches are complex chains of sugar molecules. In general, starches are broken down into simple sugar units during digestion. During **respiration**, the organism "burns" sugar molecules to release stored energy. Energy from respiration gives the cell the power to do work. There are about four calories of energy in one gram of carbohydrate. One "small" **calorie** is the energy needed to raise the temperature of one gram of water one degree Celsius. One **Food Calorie** is equal to 1,000 "small" calories.

Fats insulate and protect body organs. They comprise the "soapy" film that comprises cell membranes. Fats are also a secondary source of energy. Fats are found in nuts, butter, cheese, and meat.

Proteins are the "building blocks" of life. They are made of long **amino acid** chains that are linked according to the instructions of **DNA** molecules in the nucleus of body cells. Proteins make up the structure of our skin, hair, muscle tissue, and more. There are eight essential amino acids found in food. Red meat, fish, poultry, dairy products, and eggs are sources of **complete proteins** which contain all the necessary amino acids. Plants such as rice, cereal, and vegetables are **incomplete proteins** that are missing one or more of the essential amino acids. Proteins must be broken down into their individual amino acids by digestion before the cells of the body can put them to good use. In addition, proteins can be burned for their energy if carbohydrates and fats are unavailable. One gram of protein releases approximately four calories of energy.

The body also needs *vitamins* and *minerals*. **Vitamins** act as "coenzymes" in the regulation of chemical reactions involved in growth and normal body functioning. **Vitamin A** (a.k.a. retinol) helps to maintain healthy eyes, skin, and the linings of the digestive tract. **Vitamin D** (a.k.a. cholecalciferol) helps to maintain healthy bones. Vitamin D deficiency results in poorly formed bones: a condition called "rickets." **Vitamin E** (a.k.a. alpha-tocopherol) protects cell parts from damage by oxygen. Vitamin E deficiency results in the degeneration of nerve and muscle tissue. **Vitamin K** is needed for healthy blood clotting in the event of injury to the body. **Vitamins A, D, E,** and **K** are "fat soluble" vitamins stored in fat tissue. **Vitamin C** and **Vitamin B complex** are "water soluble" vitamins that cannot be stored in the body. Excess vitamins C and vitamin B complex are washed out of the body and must be ingested on a daily basis. Vitamin C is needed for the production of the body's energy reserves. Vitamin C deficiency results in "scurvy." Symptoms of scurvy include bleeding gums, swollen joints, skin bruises, and overall weakness. Vitamin B complex consists of several related vitamins needed for general health maintenance. **Minerals** such as the ions of **sodium, chlorine, calcium,** and **iodine** are also needed for the maintenance of a healthy body. Like vitamins, minerals help regulate **cellular metabolism.**

     Water provides the "medium" in which all the body's chemical reactions can take place. This is why water is so essential for life. Blood plasma—which carries all of the nutrients throughout the bloodstream to the body's individual cells—is 92 percent water. Water also helps the body to maintain a regular temperature of about 98.6° Fahrenheit (about 37° Celsius). This is because water stores heat energy better than most other substances.

Homework Directions

Make a calendar chart to keep a record of all the meals you eat this week including snacks. Next to each food item you eat, including fluids, list the nutrients (i.e., carbohydrates, fats, amino acids, proteins) you are sure are in that food. Compare your daily diet to the Chart on Journal Sheet #1. Write a paragraph of about 50 words that honestly summarizes your good and bad eating habits.

Assignment due: _____

_____   _____   ____/____/____
     Student's Signature   Parent's Signature   Date

NUTRITION CHEMISTRY

Work Date: ____/____/____

LESSON OBJECTIVE

Students will keep a daily record of the foods they eat and analyze their nutritional value.

Classroom Activities

On Your Mark!

Give students 10 minutes to read their Fact Sheet. Discuss and list the primary function of each of the nutrients described in the Fact Sheet and have students copy that information on Journal Sheet #1. Have them describe some of the consequences of having a poor diet.

Get Set!

Assist students in creating a table on Journal Sheet #1 that they can use in completing the Homework Assignment. Their table should reflect the standard "Nutrition Facts" label that appears by law on every food package sold as shown in Table A. Students will need to estimate the amounts of preprepared foods they eat at fast food restaurants or at home. They can do this by remembering that there are 2.2 pounds in one kilogram. A large hamburger (i.e., a quarter pounder), therefore, contains about 115 grams of food, although not all of it is pure protein. Each gram of food, containing carbohydrate or protein, yields about 4 Food calories. Point out that thiamine, riboflavin, and niacin are vitamins B_1, B_2, and B_3, respectively.

	food item	Food calories from carbohydrate and fat	Protein (grams)	% Vitamin A	% Thiamine	% Riboflavin	% Niacin	% Vitamin C
TABLE A: NUTRITIONAL VALUE OF MY DIET								
Day #1								
Day #2								
Day #3								

Go!

Give students ample time to examine Nutrition Facts labels and set up their table.

Materials

Journal Sheet #1, packaged food items with standard "Nutrition Facts" labels

CH13 JOURNAL SHEET #1

NUTRITION CHEMISTRY

		RECOMMENDED DAILY ALLOWANCE OF NUTRIENTS									
	age	weight (kg)†	Food calories	protein (g)	Vitamin A*	Vitamin B$_1$	Vitamin B$_2$	Vitamin B$_3$	Vitamin C	Vitamin D	Vitamin E
males	10-12	35	2500	45	4500	1.3	1.3	17	40	400	20
	12-14	43	2700	50	5000	1.4	1.4	18	45	400	20
	14-18	59	3000	60	5000	1.5	1.5	20	55	400	25
females	10-12	35	2250	50	4500	1.1	1.3	15	40	400	20
	12-14	44	2300	50	5000	1.2	1.4	15	45	400	20
	14-16	52	2400	55	5000	1.2	1.2	16	50	400	25
	16-18	54	2300	55	5000	1.2	1.5	15	50	400	25

* The amounts of vitamins A, D, and E are given in International Units. The amounts of vitamins B$_1$, B$_2$, B$_3$, and C are given in milligrams.

† To obtain your weight in kilograms divide your weight in pounds by 2.2. There are 2.2 pounds in one kilogram.

NUTRITION CHEMISTRY

Work Date: _____/_____/_____

LESSON OBJECTIVE

Students will test a food sample for the presence of protein.

Classroom Activities

On Your Mark!

Before the start of class prepare a 3 percent copper sulphate ($CuSO_4$) solution by dissolving 3 grams of copper sulphate in 97 ml of water. Prepare a 10 percent potassium hydroxide (KOH) solution by dissolving 10 grams of potassium hydroxide in 97 ml of warm water on a hot plate. Keep the two solutions separate until used. Set aside samples of lean ground beef, egg white, margarine, and plain sugar.

Begin with a review of the structure of proteins and point out that strong alkalies such as potassium hydroxide turn blue-violet in the presence of the peptide bonds that link amino acids together and copper sulphate.

Get Set!

Discuss the procedure used to identify which food samples contain protein in the experiment described in Figure A on Journal Sheet #2. Have students create their own table on Journal Sheets #2 to record their observations.

Go!

Assist students in completing the experiment described in Figure A on Journal Sheet #2. When they are finished give them time to review their Homework Assignment to date which requires a daily reporting of the nutritional value of yesterday's "menu."

Materials

hot plate, balance, copper sulphate, potassium hydroxide, water, beakers, medicine droppers, lean ground beef, lard or margarine, eggs, sugar, petri dishes

CH13 JOURNAL SHEET #2

NUTRITION CHEMISTRY

FIGURE A

<u>Directions</u>: (1) Place several grams (i.e., a teaspoon) of lean ground beef, lard or margarine, egg white, and sugar in 4 separate petri dishes. (2) Mix the 3 percent copper sulphate and 10 percent potassium hydroxide solutions given to you by your instructor together in a small beaker. (3) Use a medicine dropper to place several drops of the light blue mixture onto each sample. (4) Record your observations. The mixed reagent called Biuret reagent turns blue-violet in the presence of protein.

GENERAL SAFETY PRECAUTIONS

Wear goggles. Avoid direct skin contact with the food samples or the chemical solutions. Uncooked food contains bacteria. The chemical solutions used are caustic.

NUTRITION CHEMISTRY

Work Date: ____/____/____

LESSON OBJECTIVE

Students will test a food sample for the presence of starch.

Classroom Activities

On Your Mark!

Before the start of class prepare a beaker of Lugol's solution. Dissolve 5 grams of potassium iodide (KI) in 1,000 ml of water. Add 2 grams of resublimed iodine crystals or tincture of iodine. Set aside samples of lean ground beef, bread, egg white and corn starch.

Begin with a description of **starch**. Explain that starch is a carbohydrate composed of linked sugars molecules. In the presence of starch, Lugol's solution turns light blue or green. If time permits, students can boil the bread and meat samples before they place them in their petri dishes. Warmed starch changes color more quickly than raw starch.

Get Set!

Discuss the procedure used to identify which food samples contain starch in the experiment described in Figure B on Journal Sheet #3. Have students create their own table on Journal Sheet #3 to record their observations.

Go!

Assist students in completing the experiment described in Figure B on Journal Sheet #3. When they are finished give them time to review their Homework Assignment to date which requires a daily reporting of the nutritional value of yesterday's "menu."

Materials

hot plate, balance, petri dishes, potassium iodide, resublimed iodine crystals or tincture of iodine, water, beakers, medicine droppers, lean ground beef, bread, eggs, corn starch

CH13 JOURNAL SHEET #3

NUTRITION CHEMISTRY

FIGURE B

Directions: (1) Place several grams (i.e., a teaspoon) of lean ground beef, bread, egg white, and corn starch in 4 separate petri dishes. (2) Use a medicine dropper to place several drops of the solution given to you by your instructor on each sample. (3) Record your observations. The reagent called Lugol's reagent turns light blue or green in the presence of starch.

GENERAL SAFETY PRECAUTIONS

Wear goggles. Avoid direct skin contact with the food samples or the chemical solutions. Uncooked food contains bacteria and the chemical solutions used are caustic.

NUTRITION CHEMISTRY

Work Date: ____/____/____

LESSON OBJECTIVE

Students will convert starch to simple sugar.

Classroom Activities

On Your Mark!

Benedict's solution is a sodium citrate, sodium carbonate, copper sulphate solution that can be purchased from any laboratory supply house. In the presence of glucose, the solution turns an orange or yellowish-red color.

Begin the lesson by reviewing the composition of **starch**. Explain that Benedict's solution turns an orange or yellowish-red color. It is an indicator of simple sugars.

Get Set!

Discuss the procedure used to identify sugar in the experiment described in Figure C on Journal Sheet #4. Be sure students know exactly how to handle *Hot Test Tubes* with heat resistant gloves or tongs. Warn them to avoid inhaling any of the gases or hot vapors (i.e., steam) produced by the experiment.

Go!

Assist students in completing the experiment described in Figure C on Journal Sheet #4. When they are finished give them time to review their Homework Assignment to date which requires a daily reporting of the nutritional value of yesterday's "menu."

Materials

Bunsen burners, matches, ring stands and clamps, large test tubes, test tube tongs, sodium carbonate pellets, concentrated hydrochloric acid (i.e., 37 percent), Benedict's solution, water, corn starch

CH13 Journal Sheet #4

NUTRITION CHEMISTRY

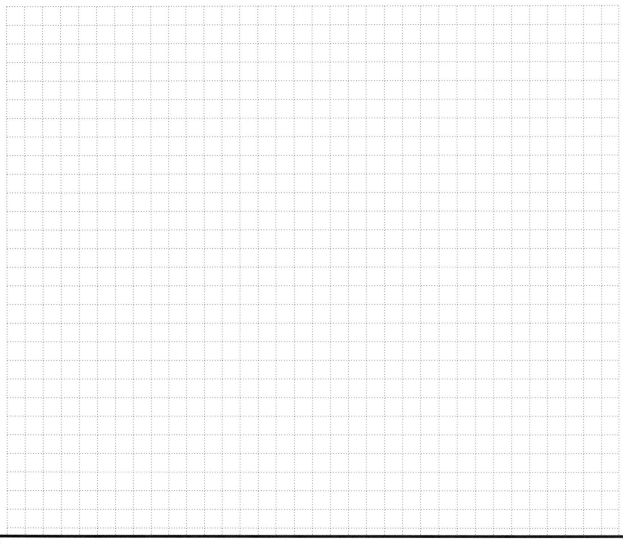

FIGURE C

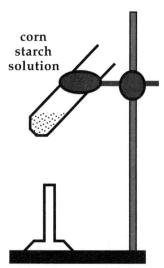

Directions: (1) Pour 5 ml of water into a large test tube. (2) Place a gram (i.e., a "pinch") of corn starch into the water. (3) Light the Bunsen burner and ADJUST TO A SMALL FLAME. (4) Heat the test tube to a SLOW BOIL, then turn off the flame. (5) Have your instructor add one drop of concentrated hydrochloric acid to your test tube. (6) Use a test tube holder to transfer your test tube to the sink. (7) Hold the test tube over the sink and add several pellets of the sodium carbonate provided by your instructor until the bubbles disappear. (8) Add a drop of the Benedict's solution provided by your instructor to the test tube. (9) Secure the test tube over the Bunsen burner, light the burner, and reheat over a small flame. (10) Record your observations. Benedict's solution turns orange (or yellow-red) in the presence of simple sugar (i.e., glucose).

corn starch solution

GENERAL SAFETY PRECAUTIONS

Wear goggles and use heat-resistant gloves or tongs. Use extreme caution in the handling of equipment that is hot.

CH13 Review Quiz

Directions: Keep your eyes on your own work.
Read all directions and questions carefully.
THINK BEFORE YOU ANSWER!
Watch your spelling, be neat, and do the best you can.

CLASSWORK   (~40): _____
HOMEWORK   (~20): _____
CURRENT EVENT   (~10): _____
TEST   (~30): _____

TOTAL   (~100): _____
(A ≥ 90, B ≥ 80, C ≥ 70, D ≥ 60, F < 60)

LETTER GRADE: _____

TEACHER'S COMMENTS: _____

_____

NUTRITION CHEMISTRY

LIST THREE (3) MAIN FUNCTIONS OF FOOD. *9 points*

(1) _____   (2) _____   (3) _____

TRUE–FALSE FILL-IN: If the statement is true, write the word TRUE. If the statement is false, change the underlined word to make the statement true. *15 points*

_____ 4. <u>Nutrients</u> are the usable substances in food.

_____ 5. <u>Carbohydrates</u> are an organism's primary source of energy.

_____ 6. Sugars and starches are the two main types of <u>minerals</u>.

_____ 7. Plants produce sugar during <u>photosynthesis</u> by trapping the radiant energy of the sun.

_____ 8. During <u>respiration</u>, an organism "burns" sugar molecules to release stored energy.

_____ 9. One "small" calorie is the energy needed to raise the temperature of one gram of <u>sugar</u> one degree Celsius.

_____ 10. One Food Calorie is equal to <u>10,000</u> "small" calories.

_____ 11. <u>Proteins</u> are a secondary source of energy found in nuts, butter, cheese, and meat.

_____ 12. <u>Vitamins</u> are long amino acid chains that link together according to the instructions of the DNA molecules.

_____ 13. Red meat, fish, poultry, dairy products, and eggs are sources of <u>incomplete</u> proteins which contain all the necessary amino acids.

_____ 14. Plants such as rice, cereal, and vegetables are <u>complete</u> proteins that are missing one or more of the essential amino acids.

_____ 15. Vitamins A, D, and E are "fat soluble" vitamins stored in <u>muscle</u> tissue.

_____ 16. Vitamins C and B are "water soluble" vitamins that <u>cannot</u> be stored in the body.

CH13 Review Quiz (cont'd)

_____17. <u>Vitamins</u> like the ions of sodium, chlorine, and calcium are also needed for the maintenance of a healthy body.

_____18. Blood plasma, which carries all of the nutrients throughout the blood-stream to the body's individual cells, is 92 percent <u>water</u>.

MATCHING: Choose the letter of the phrase on the right that describes the vitamin on the left. *6 points*

_____ 19. vitamin A   (A)   helps to maintain healthy eyes and skin

_____ 20. vitamin B   (B)   helps blood to clot

_____ 21. vitamin C   (C)   needed for the production of energy

_____ 22. vitamin D   (D)   protects cells from damage by oxygen

_____ 23. vitamin E   (E)   helps to maintain healthy bones

_____ 24. vitamin K   (F)   a complex of several vitamins

_____   _____   ____/____/____
Student's Signature   Parent's Signature   Date

CHEMICAL CYCLES

TEACHER'S CLASSWORK AGENDA AND CONTENT NOTES

Classwork Agenda for the Week

1. Students will identify the limited chemical resources we use every day and list ways they can be conserved for future use.

2. Students will show how the environment recycles water.

3. Students will show how carbon dioxide can be recycled in the environment.

4. Students will describe how nitrogen is recycled throughout the environment and summarize other cycles related to that process.

Content Notes for Lecture and Discussion

The Greek mathematician and scientist **Thales** (b. 624 B.C.; d. 547 B.C.) was one of the first pre-Socratic philosophers to suggest that water circulated from rivers, streams, lakes, and oceans to the atmosphere and back. He was the first to propose the existence of a **hydrologic cycle**. The idea that specific and immutable chemical elements and compounds are "reused" by nature simplified the study of chemistry. Before Thales, early alchemists harbored a more "mystical" view of chemical interactions. The alchemists believed that chemistry could be used to transmute "temporal substances" into spiritually perfect, immortal forms. They sought to create "elixirs" that would prolong life and produce enlightenment; and, their experiments were based more on trial and error than any systematic approach. The simple notion that chemical substances went through cycles, however, gave chemists a starting point from which to examine chemical reactions.

The German chemist **Georg Ernst Stahl** (b. 1660; d. 1734) proposed that plants and animals recycled "phlogiston," a combustible substance that allowed materials to burn. While the work of **Antoine Laurent Lavoisier** (b. 1743; d. 1794) proved that phlogiston did not exist, Stahl's suggestion that chemical cycles took place between plants and animals led to the discovery of the **carbon** and **oxygen cycles**. The work of English chemist **Joseph Priestley** (b. 1733; d. 1804), Dutch plant physiologist **Jan Ingenhousz** (b. 1730; d. 1799), and the Swiss botanist **Jean Senebier** (b. 1742; d. 1809) laid the foundation for the work of later chemists. The German chemist **Julius von Sachs** (b. 1832; d. 1897) showed that **starch** present in plant cells was the product carbon dioxide absorbed from the atmosphere. And, the German botanist **Andreas Franz Wilhelm Schimper** (b. 1856; d. 1901) demonstrated that solar energy could be "trapped and stored" by plants for later use by animals.

Nineteenth century farmers were well aware that nitrogen was essential to the growth of plants. However, no one was sure where the nitrogen originated nor how it was absorbed. Most chemists agreed that the most likely source of nitrogen was the atmosphere which was nearly four-fifths nitrogen. The Swiss geologist **Horace Bénédict de Saussure** (b. 1740; d. 1799) concluded that nitrogen was taken up by the roots but the actual mechanism remained controversial for fifty years. Others believed the gas was absorbed directly from the atmosphere by the leaves. Saussure's son, the plant physiologist **Nicholas Théodore de Saussure** (b. 1767; d. 1845) was a notable botanist in his own right, contributing much to the elucidation of the carbon-oxygen cycle. The French agricultural chemist **Jean-Baptiste Boussingault** (b. 1802; d. 1887) finally proved that plants thrived in soil mixtures rich in ammonia salts and nitrates regardless of the controlled contents of the surrounding atmosphere. In 1862, the French microbiologist **Louis Pasteur** (b. 1822; d. 1895) proposed that microorganisms might be involved in the plant absorption of nitrogen-containing salts. But it wasn't until 1887 that bacteria living on the roots of **leguminous plants** were found to convert diatomic nitrogen (N$_2$) to ammonia (NH$_3$). Legumes are plants belonging to the pea family having pods and dry fruits. The bacteria thriving in symbiosis with the legume is generally of the genus *Rhizopus*.

In Lesson #1, students will identify the limited chemical resources we use every day and list ways they can be conserved for future use.

In Lesson #2, students will show how the environment recycles water.

In Lesson #3, students will show how carbon and oxygen can be recycled through the environment.

In Lesson #4, students will describe how nitrogen is recycled throughout the environment and summarize other cycles related to that process. As an alternative to this activity, you may choose to crush the nodules attached to the roots of leguminous plants (i.e., alfalfa, clover, and peanuts) onto a slide and view the nitrogen-fixing bacteria contained in the nodules under high power. The bacteria can also be seen by placing a drop of water containing the bacteria onto a slide smeared with dried methylene blue stain.

ANSWERS TO THE HOMEWORK PROBLEMS

1. Answers will vary but should express the notion that chemical elements and compounds are limited resources and that life must use and reuse essential materials.

2. Answers will vary but should express the notion that pollutants are chemicals that can interfere with natural chemical cycles.

3. Answers will vary but should express the notion that insecticides are chemicals that can interfere with natural chemical cycles.

ANSWERS TO THE END-OF-THE-WEEK REVIEW QUIZ

1. limited	6. C	BRIEF ESSAYS: Students' essays should reflect their study
2. true	7. B	of the FACT SHEET which clearly summarizes each
3. recycles	8. D	chemical cycle.
4. true	9. A	
5. true		

CH14 Fact Sheet

CHEMICAL CYCLES

CLASSWORK AGENDA FOR THE WEEK

(1) Identify the limited chemical resources we use every day and list ways they can be conserved for future use.
(2) Show how the environment recycles water.
(3) Show how carbon and oxygen can be recycled.
(4) Describe how nitrogen is recycled throughout the environment.

The earth has limited chemical resources. Many of us are conscious of this fact when we "recycle" materials like newspaper and aluminum cans. We recycle newspaper to help conserve trees. Trees perform chemical reactions like **photosynthesis** that help to insure the survival of all living things on our planet. We recycle aluminum cans to help conserve the landscape. If we simply buried all of the cans we use in waste dumps we would need to destroy the landscape in search of replacement metals. Nature recycles chemical resources as well. And, it is important to remember that **pollution** interferes with nature's ability to recycle natural resources.

A **cycle** is a set of events that are repeated over and over again. In nature, there are four basic chemical cycles: the **water cycle**, the **carbon cycle**, the **oxygen cycle**, and the **nitrogen cycle**.

There are more than 2 billion trillion (2×10^{21}) liters of **water** in the oceans, lakes, rivers, streams, polar ice caps, and atmosphere covering the earth. And all of it is recycled by physical means. Warming the oceans, lakes, rivers, and streams of the world causes water to evaporate into the atmosphere. In the atmosphere, water vapor cools and condenses into rain, frost, or snow. It is pulled back to earth by the force of gravity only to be evaporated again and again. In the process, water is cleansed of contaminating impurities.

Carbon and **oxygen** are recycled by chemical means. During photosynthesis, plants use the energy of the sun to "fix" the carbon in carbon dioxide molecules into energy rich sugar. Oxygen is given off in the process. All living organisms use sugar and oxygen during **respiration** to satisfy their energy needs. In the process, carbon dioxide is returned to the atmosphere.

Nitrogen is also recycled by chemical means. About 79 percent of our atmosphere is composed of "free" **diatomic nitrogen** (N_2). During lightning storms, nitrogen is mixed with water vapor to form **nitrites** (NO_2^-) and **nitric acid** (HNO_3). Rain brings this "fixed" nitrogen to the ground where it is absorbed by bacteria in the soil. **Nitrogen-fixing bacteria** provide plants with the nitrogen they need to make proteins and nucleic acids: two of the basic molecules of life. Other forms of bacteria called **decomposers** break down these macromolecules in decaying matter buried in the soil and return diatomic nitrogen, a light gas, to the atmosphere.

Pollution refers to those substances that interfere with the natural recycling of earth's raw materials. **Chemical pollutants** are the byproducts of many of modern industry's manufacturing processes. When pollutants escape into the environment natural chemical cycles are upset. One of the fastest growing fields in science today is the field of **ecology**. Ecologists study how living things on our planet depend on one another and the nonliving matter in our environment. Our failure to learn as much as we can about "environmental chemistry" could one day result in the extinction of the human race.

Homework Directions

Directions: Write a short paragraph that clearly answers each of the following questions.

1. Why are cycles important in nature?
2. How can soot-filled smoke from industrial plants affect chemical cycles in nature?
3. How might insecticides sprayed on plants to help growing crops affect chemical cycles in nature?

Assignment due: _____

_____	_____	____/____/____
Student's Signature	Parent's Signature	Date

CHEMICAL CYCLES

Work Date: ____/____/____

LESSON OBJECTIVE

Students will identify the limited chemical resources we use every day and list ways they can be conserved for future use.

Classroom Activities

On Your Mark!

Ask students to explain why the earth has limited natural resources. The obvious answer is that the earth is "an island in space." It is a "self-contained system" in which chemical substances react and rereact in set patterns of chemical activity. Individual chemical elements and compounds must be used and reused in all sorts of chemical reactions because no new matter can ever be created or destroyed according to the **Law of Conservation of Matter**. Explain that any set of chemical events that is repeated over and over again is called a **cycle**. List the four basic chemical cycles that occur in nature and have students copy your list onto Journal Sheet #1: the **water cycle**, the **carbon cycle**, the **oxygen cycle**, the **nitrogen cycle**. An intriguing idea about cycles was proposed by the mathematician **John Allen Paulos** in his book *Innumeracy* in 1988. He asked: "What is the probability that you are inhaling molecules of carbon dioxide exhaled by the Roman leader Julius Caesar at his dying breath?" Assuming that it takes exhaled molecules about two thousand years to be thoroughly spread around the earth's atmosphere, the odds are quite surprising. Refer to Paulos's calculations to see that there is a 99% probability that you did indeed inhale a molecule of a carbon dioxide once exhaled by the infamous Roman general. Truly, we may all eventually "become part of one another."

Get Set!

Tell students to brainstorm a list of "recyclable" products they use in their everyday lives. Have them choose one of those products and discuss in detail how the product and its components might be recycled. Instruct them to use their imagination to come up with a logical sequence of processes that people might use to make their product available for use after it is "discarded." Have them record their ideas on Journal Sheet #1. Draw Illustration A to give them the idea.

ILLUSTRATION A

natural or recycled resources

discarded product treated to make it "resourceful"

manufactured product

used product discarded

product packaged and sold

Go!

Distribute large sheets of butcher paper and have students draw the steps their product goes through in its "recycled lifetime."

Materials

butcher paper, crayons or colored marking pens

CH14 JOURNAL SHEET #1

CHEMICAL CYCLES

CH14 Lesson #2

CHEMICAL CYCLES

Work Date: ____/____/____

LESSON OBJECTIVE

Students will show how the environment recycles water.

Classroom Activities

On Your Mark!

Prepare the bent glass tubing shown in Figure A on Journal Sheet #2 before the start of class. Prepare a solution of "sea water" by mixing salt with sand or muddy pond water. Set the mixture aside in a large jar or beaker.

Begin class discussion with a review of the **phases of matter**. Give students time to recall and discuss how solids are melted to form liquids, how liquids are vaporized to form gases, how vapors are condensed by cooling to form liquids again, and how liquids can be frozen to reform solids. The phases of matter can change from one form to another in a "natural cycle." **Water** is recycled and purified in such a way by the warming effects of solar radiation and the cooling effects of rising elevation and decreasing

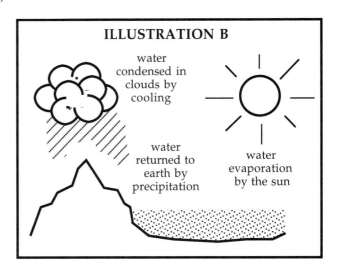

ILLUSTRATION B

water condensed in clouds by cooling

water returned to earth by precipitation

water evaporation by the sun

atmospheric pressure. Draw Illustration B on the board and have students copy the illustration on Journal Sheet #2.

Get Set!

Remind students that they performed a similar demonstration in Lesson #3 in the *Mixtures* unit. In that demonstration they separated the components of a solution by distillation. In this demonstration they will separate "sea water" and show how fresh water is returned to the "sea."

Go!

Give students ample time to set up and perform the demonstration described in Figure A on Journal Sheet #2.

Materials

ring stand and clamps, Ehrlenmeyer flasks, single-holed rubber stoppers, glass tubing, small beakers, Bunsen burners, "sea water" mixture, matches

CH14 Journal Sheet #2

CHEMICAL CYCLES

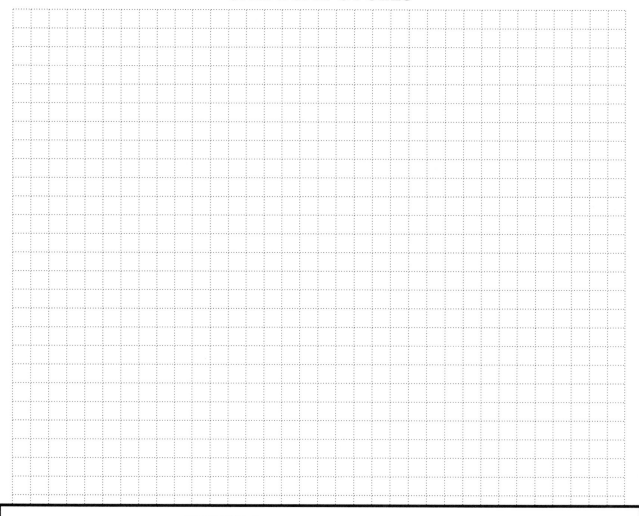

FIGURE A

Directions: (1) Pour 100 milliliters of the "sea water" mixture given to you by your instructor into an Ehrlenmeyer flask. (2) Pour about 50 ml of the same solution into a small beaker. (3) Place the flask on the ring stand and secure it with a clamp so that it cannot be toppled. (4) Place the beaker next to the ring stand. (5) Insert the rubber stopper holding the glass tubing snugly into the flask. DO NOT HOLD OR MANIPULATE THE GLASS TUBING. HANDLE THE RUBBER STOPPER ONLY. Make sure the other end of the glass tubing is positioned over the beaker. (6) Turn on the Bunsen burner. (7) Record what you observe for the next ten minutes and explain your observations. What is the color and consistency of the evaporated water compared to the "sea water?" What process allows the pure water to return to the "sea"?

GENERAL SAFETY PRECAUTIONS

Be sure you are familiar with the proper use of a Bunsen burner. Heat solutions and mixtures slowly. Wear goggles to protect your skin and eyes from being burned by SCALDING HOT STEAM. Do not touch any part of the equipment without heat-resistant gloves or tongs. Clean up when the apparatus is cool.

CHEMICAL CYCLES

Work Date: ____/____/____

LESSON OBJECTIVE

Students will show how carbon dioxide can be recycled in the environment.

Classroom Activities

On Your Mark!

Prepare a solution of **limewater** by adding an excess of calcium hydroxide $[(Ca(OH)_2]$ or calcium oxide (CaO) to distilled water until the solution is saturated. Allow the solution to sit overnight, then pour off or filter the supernatant (i.e., the liquid above the precipitate) into a jar or large flask. Cap off the clear solution and set it aside. Preprepared solutions of limewater can also be purchased commercially.

Begin class discussion with a review of the chemical equations for **photosynthesis** and **respiration**. These chemical reactions comprise a cycle that uses and reuses carbon dioxide (CO_2) and oxygen (O_2). The **carbon-oxygen cycle** depends upon a variety of chemical catalysts (i.e., **plant chlorophyll**) and, of course, on the availability of solar radiation, plants, and animals. Plants and animals are, therefore, dependent upon one another for survival.

PHOTOSYNTHESIS: $\quad 6CO_2 \quad + \quad 6H_2O \quad \rightarrow \quad C_6H_{12}O_6 \quad + \quad 6O_2$

RESPIRATION: $\quad\quad\quad 6O_2 \quad + \quad C_6H_{12}O_6 \quad \rightarrow \quad 6CO_2 \quad + \quad 6H_2O$

Explain that carbon dioxide (CO_2) is also used by sea animals to manufacture their hard outer shells which are made of calcium carbonate ($CaCO_3$).

Get Set!

Have students copy the following sequence of chemical equations on Journal Sheet #3.

(A) $\quad CO_2 \quad + \quad H_2O \quad \rightarrow \quad H_2CO_3$

(B) $\quad Ca(OH)_2 \quad + \quad H_2CO_3 \quad \rightarrow \quad CaCO_3 \quad + \quad 2H_2O$

(C) $\quad CaCO_3 \quad + \quad 2HCl \quad \rightarrow \quad CaCl_2 \quad + \quad H_2O \quad + \quad CO_2$

Explain that exhaled carbon dioxide can combine with water to produce carbonic acid (H_2CO_3). Carbonic acid reacts with calcium hydroxide to produce calcium carbonate [i.e., chalk and sea shells ($CaCO_3$)]. Most any acid [i.e., hydrochloric acid (HCl)] reacts with calcium carbonate to produce a salt called calcium chloride ($CaCl_2$), water and carbon dioxide.

Go!

Assist students in performing the experiment described in Figure B on Journal Sheet #3. Have them explain why this series of chemical reactions comprises a cycle of chemical events.

Materials

limewater or calcium hydroxide/calcium oxide and water for preparation, Ehrlenmeyer flasks, rubber tubing, filters and funnels, beakers, mild hydrochloric acid (1 molar solution available through chemical supply houses)

CH14 JOURNAL SHEET #3

CHEMICAL CYCLES

FIGURE B

<u>Directions</u>: (1) Pour 100 milliliters of the limewater solution given to you by your instructor into an Ehrlenmeyer flask. (2) Insert a length of rubber tubing into the flask and insert your personal straw into the tubing. (3) Exhale several times into the rubber tubing and allow your groupmates to do the same. (4) Record your observations. (5) Place a glass funnel into a second Ehrlenmeyer flask and place a folded piece of filter paper into the funnel. (6) Remove the rubber tubing from the first flask and carefully pour the solution from the first flask through the funnel. (7) Set the damp filter paper aside in a dry place and leave it over-night. (8) When you return the next day, test a drop of the mild hydrochloric acid solution given to you by your instructor on the dried remains on the filter paper. (9) Record your observations. What gas do you think was produced?

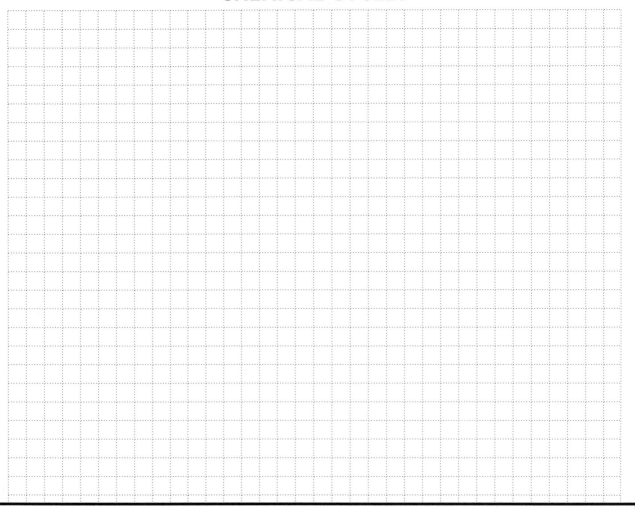

exhale into tubing

filter milky solution

test acid drop on dried crystals

GENERAL SAFETY PRECAUTIONS

Wear goggles to protect your eyes against splattered lime-water and acid. Avoid contact with the substances in this demonstration. They are caustic.

CH14 Lesson #4

CHEMICAL CYCLES

Work Date: ____/____/____

LESSON OBJECTIVE

Students will describe how nitrogen is recycled throughout the environment and summarize other cycles related to that process.

Classroom Activities

On Your Mark!

Give students several minutes to complete the experiment started in Lesson #3. Discuss and explain their results. The set of chemical reactions recycled carbon dioxide (i.e., from their lungs, into the calcium carbonate, and back to the atmosphere).

Draw the diagram shown in Illustration C and have students copy it on Journal Sheet #4. Explain how nitrogen is recycled throughout the environment using the information in the Teacher's Agenda and Content Notes. Identify the roles of primary producers (land and water plants that make sugar), primary and secondary consumers (herbivorous zooplankton and carnivorous insects and fish, respectively), and decomposers (bacteria) in an ecosystem.

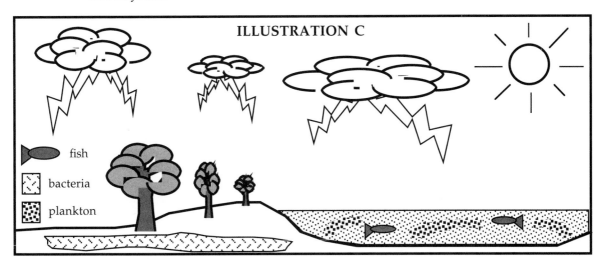

ILLUSTRATION C

fish

bacteria

plankton

Get Set!

Instruct students to transfer the diagram onto a large sheet of butcher paper and complete the illustration with explanations and arrows that describe the relationships among the different groups of plants, animals, and natural resources. Their final illustration should trace the pathways of the elements and compounds of the four basic natural chemical cycles.

Go!

Give students ample time to design and complete their poster.

Materials

butcher paper, crayons

CH14 JOURNAL SHEET #4

CHEMICAL CYCLES

CH14 REVIEW QUIZ

Directions: Keep your eyes on your own work.
Read all directions and questions carefully.
THINK BEFORE YOU ANSWER!
Watch your spelling, be neat, and do the best you can.

CLASSWORK (~40): _____
HOMEWORK (~20): _____
CURRENT EVENT (~10): _____
TEST (~30): _____

TOTAL (~100): _____
(A ≥ 90, B ≥ 80, C ≥ 70, D ≥ 60, F < 60)

LETTER GRADE: _____

TEACHER'S COMMENTS: _____
_____

CHEMICAL CYCLES

TRUE–FALSE FILL-IN: If the statement is true, write the word TRUE. If the statement is false, change the underlined word to make the statement true. *10 points*

_____ 1. The earth has <u>unlimited</u> chemical resources.

_____ 2. Trees perform chemical reactions like <u>photosynthesis</u> that help to insure the survival of all living things on our planet.

_____ 3. Nature <u>creates new</u> chemical resources.

_____ 4. <u>Pollution</u> interferes with nature's ability to recycle natural resources.

_____ 5. A <u>cycle</u> is a set of events that is repeated over and over again.

MATCHIING: Choose the letter of the word or phrase that best describes each chemical cycle. *8 points*

_____ 6. water cycle   (A) involves action by bacteria

_____ 7. carbon cycle   (B) gets "fixed" in photosynthesis

_____ 8. oxygen cycle   (C) never changes chemical form

_____ 9. nitrogen cycle   (D) gets "burned" during respiration

BRIEF ESSAYS: Summarize each chemical cycle in three or four short sentences. *12 points*

THE WATER CYCLE

_____

_____

_____

_____

THE CARBON AND OXYGEN CYCLES

_____

_____

_____

_____

THE NITROGEN CYCLE

_____

_____

_____

_____

_____   _____   ____/____/____
Student's Signature   Parent's Signature   Date

NUCLEAR REACTIONS AND RADIOACTIVITY

TEACHER'S CLASSWORK AGENDA AND CONTENT NOTES

Classwork Agenda for the Week

1. Students will draw models of the structure of the atomic nuclei and explain what is meant by the term "isotope."

2. Students will describe the different types of radiation and write nuclear equations.

3. Students will explain how the "half-life" of atomic isotopes is used to find the age of ancient objects.

4. Students will build models of atomic nuclei and illustrate the difference between fission and fusion.

Content Notes for Lecture and Discussion

The study of **radioactivity** could not have begun without the invention of **photography**. Radioactive particles are invisible to the naked eye and their detection requires their interaction with chemical substances (i.e., in photographic emulsions or cloud chambers) that give off light upon exposure to radioactive particles. Today, physicists use sophisticated electronic detectors and computers to register the presence of subatomic particles in laboratories around the world in **cyclotrons** and **particle accelerators**. The first important experiments done with **photochemicals** were performed by the English scientist **Elizabeth Fulhame** who published *An Essay on Combustion with a View to a New Art in Dying and Painting* in 1794. Little is known of this remarkable woman who impressed the English chemist **Joseph Priestley** (b. 1733; d. 1804) with her skill at making photographic impressions in cloth treated with silver and gold salts. In the 1830s, the French designer and inventor **Louis Daguerre** (b. 1769; d. 1851) perfected the first reliable method of fixing images on copper plates. Daguerre's photographic technique, like the **cloud chamber** invented by the Scottish physicist **Charles Thomson Rees Wilson** (b. 1869; d. 1959) and the **bubble chamber** invented in 1952 by American physicist **Donald Arthur Glaser** (b. 1926), became an invaluable tool among **nuclear physicists**.

The discovery of radioactivity was accidental. In 1896, the French physicist **Henri Becquerel** (b. 1852; d. 1908) wrapped a piece of uranium ore in dark paper and left it in a closed drawer with some photographic film. Upon developing the film for another purpose, he discovered blurry images on his photographs indicating that the film had been exposed to light. Upon further investigation, Becquerel concluded that the "light" was being emitted by the uranium ore. In 1898, the Polish-born physicist Manya Sklodowska (a.k.a. **Marie Curie**: b. 1867; d. 1934) discovered two new radioactive ores: polonium and radium. She shared the Nobel Prize in Physics for these discoveries with her husband, the French physicist **Pierre Curie** (b. 1859; d. 1906) in 1903. Madame Curie was also awarded the Nobel Prize in Chemistry in 1911. The English physicist **Ernst Rutherford** (b. 1871; d. 1937) identified **alpha** and **beta particles** during his research on the structure of the atomic nucleus. And, the German-born American physicist **Albert Einstein** (b. 1879; d. 1955) determined the amount of energy that can be released by atomic particles when they are transmuted to pure energy. Einstein's work led the Italian-born American physicist **Enrico Fermi** (b. 1901; d. 1954) to invent the first atomic reactor in Chicago in 1942. Fermi's success was a prelude to the building of the atomic bomb, a program codenamed **The Manhattan Project** and coordinated by the American physicist **Robert Oppenheimer** (b. 1904; d. 1967).

CH15 Content Notes (cont'd)

The first particle accelerator was built by the English physicist **John Douglas Cockcroft** (b. 1897; d. 1967) and his associate, the Irish physicist **Ernest Walton** (b. 1903; d. 1995) in 1930. Cockcroft and Walton used a voltage multiplier to accelerate a beam of protons which they used to bombard lithium nuclei. In 1932, they were the first to succeed in splitting the atom evidenced by the **transmutation** of lithium nuclei to helium nuclei. Since the early part of this century, physicists have succeeded in identifying a host of subatomic particles and antiparticles using particle accelerators. An **antiparticle** has the same mass as its "relative" particle but opposite spin and charge. A **positron** is the antiparticle of an electron. When an antiparticle meets a particle the two annihilate one another in a burst of energy. To date, the smallest known particle which comprises both neutrons and protons are called **quarks**. There are six different quarks thought to comprise most matter: the up quark, the down quark, the top quark, the bottom quark, the strange quark, and the charm quark. **Leptons** like electrons comprise the rest of existing matter.

In Lesson #1, students will draw models of the structure of the atomic nuclei and explain what is meant by the term "isotope."

In Lesson #2, students will describe the different types of radiation and write nuclear equations to show how atoms are transmuted from one form to another.

In Lesson #3, students will explain how the "half-life" of atomic isotopes is used to find the age of ancient objects.

In Lesson #4, students will build models of atomic nuclei to illustrate the difference between fission and fusion.

ANSWERS TO THE HOMEWORK PROBLEMS

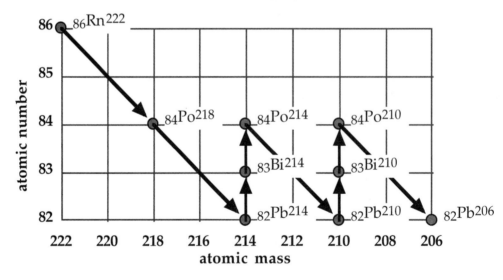

Diagonal arrows represent lost alpha particles. Vertical arrows represent lost beta particles.

ANSWERS TO THE END-OF-THE-WEEK REVIEW QUIZ

1. neutrons
2. repelled from
3. the strong force
4. true
5. true
6. true
7. sometimes
8. isotopes
9. true
10. helium
11. fusion
12. fission
13. neutron
14. $_{86}Rn^{222}$
15. $_{84}Po^{214}$
16. $_{7}N^{14}$
17. $_{90}Th^{234}$
18. $_{84}Po^{218}$

198

CH15 FACT SHEET

NUCLEAR REACTIONS AND RADIOACTIVITY

CLASSWORK AGENDA FOR THE WEEK

(1) Draw models of the structure of the atomic nuclei and explain what is meant by the term "isotope."
(2) Describe the different types of radiation and write nuclear equations.
(3) Explain how the "half-life" of atomic isotopes is used to find the age of ancient objects.
(4) Build models of atomic nuclei to illustrate the difference between fission and fusion.

The **nucleus** of every atom except hydrogen contains two kinds of **subatomic particles**: positively charged **protons** and neutral **neutrons**. One of the first questions that occurred to physicists when protons were discovered was: If protons have positive charges, and like charges repel, then the nucleus of every atom should fly apart. Why don't they? Scientists decided that atoms must be held together by a force we can call the **strong force**. Today, we know that the strong force is one thousand-trillion-trillion-trillion times (10^{39}) stronger than the force of gravity. However, it can only influence particles that are extremely close together. Otherwise, the whole universe would collapse under its power. Subatomic particles must be less than one millionth-billionth of a meter (10^{-15}m) apart in order to be glued together by the strong force. A second force that determines the composition of an atomic nucleus is called the **weak force**. The weak force is ten trillion-trillion (10^{25}) times greater than the force of gravity. But it is many times weaker than the strong force. The weak force results in the disintegration of atomic nuclei. The disintegration of atomic nuclei is called **radioactivity**.

There are two basic kinds of radioactivity: **alpha particles** and **beta particles**. Alpha particles are composed of 2 protons and 2 neutrons just like the nuclei of helium atoms. Alpha particles can be expelled from the atomic nuclei of large atoms like uranium. Losing an alpha particle changes the uranium atom to a thorium atom with two fewer protons and a mass that is reduced by 4 atomic particles. The nuclei of other atoms give off beta particles which have the same mass and charge as negatively charged electrons. Changes in the structure of the atomic nucleus are called **nuclear reactions**. **Fusion** is a process by which atomic nuclei are smashed and held together. **Fission** refers to the process by which atomic nuclei are broken apart. Scientists can write **nuclear equations** to describe these changes.

In chemical reactions all atoms of the same element react in exactly the same way. But not all atoms of the same chemical element are exactly alike. For example, some carbon atoms weigh more than other carbon atoms. Carbon nuclei all contain the same number of protons *but can have different numbers of neutrons*. Atoms with the same atomic number (i.e., number of protons) but differing atomic masses (i.e., different numbers of neutrons) are called **isotopes**. Because isotopes are radioactive, they can be identified using a device called a **Geiger counter**: a device that is sensitive to the presence of charged atomic particles. The first Geiger counter was invented by the German physicist **Hans Geiger** (b. 1882; d. 1945) in 1908. Isotopes are useful in a variety of scientific fields like **medicine** and **nuclear science**.

CH15 Fact Sheet *(cont'd)*

The rate at which atomic nuclei disintegrate is the same for atoms of the same element. That is, radioactive atoms of the same element fall apart in the same amount of time. *The time it takes for half of the atoms in an element to disintegrate is called the **half-life** of that element.* Different elements have different half-lives. Below is a list of the "half-lives" of a some **radioactive isotopes**.

RADIOACTIVE HALF-LIVES

tritium	12.26 years
carbon 14	5,730 years
oxygen 20	14 seconds
potassium 40	1,280,000,000 years
cobalt 60	5.26 years
uranium 235	710,000,000 years

As you can see, atoms of different elements disintegrate at widely different rates. Some atoms fall apart quickly while others fall apart so slowly that it is very difficult to observe their disintegration. Nevertheless, scientists can calculate the age of objects like fossils by measuring the amount of radioactive isotopes present in a sample of the object. This procedure is called **radioactive dating**.

Homework Directions

Graph the disintegration of radon-222 showing how the loss of alpha and beta particles eventually changes that isotope into lead-206. Use a graph like the one shown below to illustrate the following set of disintegrations:

$$_{86}Rn^{222} - _{84}Po^{218} - _{82}Pb^{214} - _{83}Bi^{214} - _{84}Po^{214} - _{82}Pb^{210} - _{83}Bi^{210} - _{84}Po^{210} - _{82}Pb^{206}$$

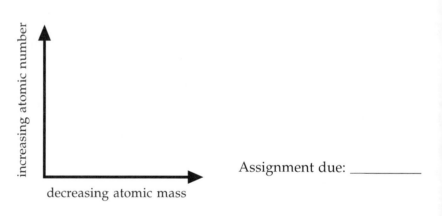

Assignment due: _____

_____   _____   ___/___/___
Student's Signature   Parent's Signature   Date

CH15 Lesson #1

NUCLEAR REACTIONS AND RADIOACTIVITY

Work Date: ____/____/____

LESSON OBJECTIVE

Students will draw models of the structure of atomic nuclei and explain what is meant by the term "isotope."

Classroom Activities

On Your Mark!

Begin the lesson wth a review of atomic structure. Draw the Bohr diagram of the carbon atom shown in Illustration A and have students recall the different particles that comprise the atom: **protons**, **neutrons**, and **electrons**. Remind students that electrons are involved in the "chemical" interactions between atoms; but explain that they are not involved in transformations of atomic nuclei. Inform students that, today, scientists have identified **quarks** as the smallest subatomic particles that comprise protons and neutrons. However, nuclear reactions can be explained in terms of the number of protons and neutrons inside the nucleus of an atom. Explain that while atoms of the same element behave exactly alike in chemical reactions, atoms of the same element do not necessarily have the same mass. That is, they contain differing numbers of neutrons. Refer students to Illustration A. Show them that carbon 14 has 2 more neutrons than carbon 12. However, both atoms have the same number of protons which determines each atom's "chemical properties." Carbon 12 and carbon 14 are **isotopes**.

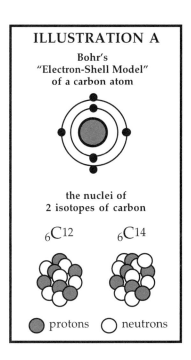

ILLUSTRATION A

Bohr's
"Electron-Shell Model"
of a carbon atom

the nuclei of
2 isotopes of carbon

$_6C^{12}$ $_6C^{14}$

● protons   ○ neutrons

Get Set!

Have students pull out their *The Periodic Table of the Elements* and review the meaning of **atomic number** and **atomic mass**. Explain that in 1815, the English physician and chemist **William Prout** (b. 1785; d. 1850) proposed that hydrogen was the basic building block of all atomic elements. He had discovered that the atomic masses of the elements were all multiples of the mass of hydrogen atoms. This assertion became known as **Prout's hypothesis**. In 1920, the New Zealand-born British physicist **Ernst Rutherford** (b. 1871; d. 1937) named the proton after William Prout since the positively charged particles inside atoms were nothing more than the nuclei of hydrogen atoms. Refer students to Table A on Journal Sheet #1 and show them how to write the symbols for different isotopes.

Go!

Give students time to discuss and complete Table A on Journal Sheet #3.

Materials

The Periodic Table of the Elements, Journal Sheet #1

CH15 JOURNAL SHEET #1

NUCLEAR REACTIONS AND RADIOACTIVITY

TABLE A					
symbol	number of protons	number of neutrons	symbol	number of protons	number of neutrons
$_1H^1$			$_1H^2$		
$_3Li^7$			$_3Li^6$		
$_6C^{12}$			$_6C^{14}$		
$_8O^{16}$			$_8O^{18}$		
$_{26}Fe^{54}$			$_{26}Fe^{57}$		
$_{50}Sn^{112}$			$_{50}Sn^{124}$		
$_{56}Ba^{137}$			$_{56}Ba^{141}$		
$_{36}Kr^{84}$			$_{36}Kr^{92}$		
$_{84}Po^{210}$			$_{84}Po^{214}$		
$_{92}U^{238}$			$_{92}U^{235}$		

NOTE: The atomic number in each symbol is written as a subscript to the lower left of each symbol. The atomic mass is written as a superscript to the upper right of each symbol.

NUCLEAR REACTIONS AND RADIOACTIVITY

Work Date: ____/____/____

LESSON OBJECTIVE

Students will describe the different types of radiation and write nuclear equations.

Classroom Activities

On Your Mark!

Begin the lesson by asking students to explain how film in a camera produces a photograph. Point out that the word "photograph" means "drawing with light." Lead the discussion to the conclusion that light particles called **photons** strike a chemically reactive emulsion that changes color by absorbing the energy of photons. Give students a brief history of photography using the information found in the Teacher's Agenda and Content Notes. Explain the importance of this "chemical art" to the study of subatomic particles by giving a brief lecture on the accidental discovery of **radioactivity** by the French physicist **Henri Becquerel** (b. 1852; d. 1908) and the discovery of **radium** by the Polish scientist **Madame Marie Curie** (b. 1867; d. 1934) and her French husband, physicist **Pierre Curie** (b. 1859; d. 1906).

Draw Illustration B on the board and instruct students to copy it on Journal Sheet #2. Explain that the properties of **alpha** and **beta particles** were "deduced" from the scattering of light flashes produced on photographic film by these **radioactive particles**. Light, negatively charged beta particles are attracted by a positive electrode producing flashes of light on the screen. Heavy, positively charged alpha particles are attracted by a negative electrode to produce similar flashes of light. Point out that the trajectory of alpha particles is less curved than that of the beta particles because the electrode upsets the pathway of the lighter particles to a greater extent than the heavier particles.

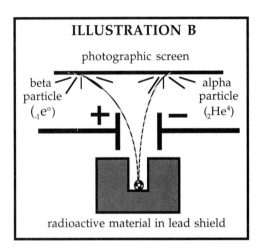

ILLUSTRATION B

photographic screen

beta particle ($_{-1}e^0$)

alpha particle ($_2He^4$)

$+$ $-$

radioactive material in lead shield

Get Set!

Inform students that changes in the contents of atomic nuclei involve **nuclear reactions**. Explain that **nuclear equations** are much like **chemical equations** in that they describe change. Have students refer to the Nuclear Equations on Journal Sheet #2. Write the Examples on the board and explain how these equations are interpreted. The loss of a beta particle is described in the first equation. This equation shows the **transmutation** of a neutron to a proton by the loss of a beta particle. This transmutation causes atoms to change to those of another element; since the atom's atomic number has increased by 1. The mass of the atom remains the same after the loss of a beta article because the particle is extremely small compared to the size of either a proton or a neutron. The loss of an alpha particle reduces an atom's atomic number by 2 resulting in its transmutation to another element. After losing an alpha particle the mass of the atom is reduced by 4 atomic particles.

Go!

Give students sufficient time to complete the remainder of the Nuclear Equations on Journal Sheet #2.

Answers to Journal Sheet Nuclear Equations—$_7N^{14}$; $_6C^{14}$; alpha; beta; $_{54}Xe^{137}$; $_{37}Rb^{92}$; alpha; alpha

Materials

Journal Sheet #2

CH15 JOURNAL SHEET #2

NUCLEAR REACTIONS AND RADIOACTIVITY

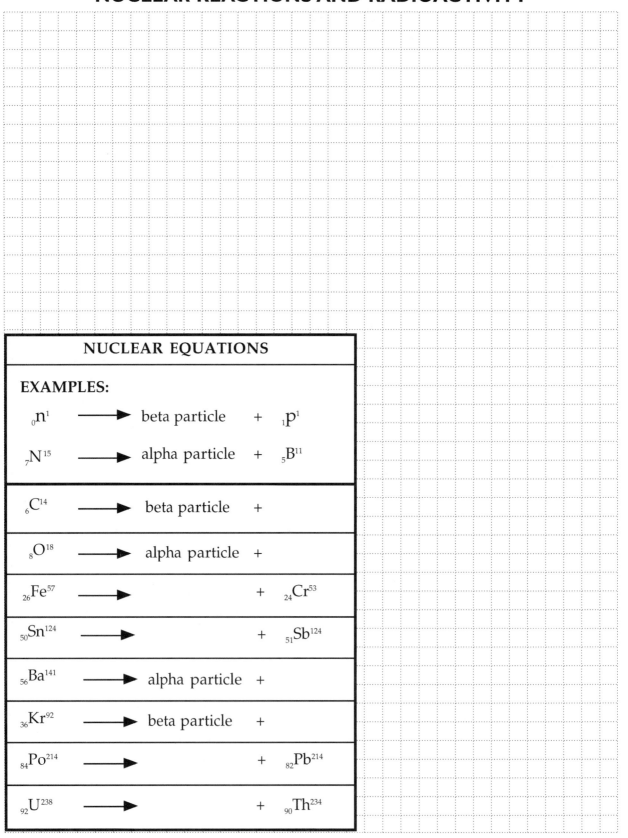

NUCLEAR EQUATIONS		
EXAMPLES:		
$_0n^1$ ⟶	beta particle +	$_1P^1$
$_7N^{15}$ ⟶	alpha particle +	$_5B^{11}$
$_6C^{14}$ ⟶	beta particle +	
$_8O^{18}$ ⟶	alpha particle +	
$_{26}Fe^{57}$ ⟶	+	$_{24}Cr^{53}$
$_{50}Sn^{124}$ ⟶	+	$_{51}Sb^{124}$
$_{56}Ba^{141}$ ⟶	alpha particle +	
$_{36}Kr^{92}$ ⟶	beta particle +	
$_{84}Po^{214}$ ⟶	+	$_{82}Pb^{214}$
$_{92}U^{238}$ ⟶	+	$_{90}Th^{234}$

NUCLEAR REACTIONS AND RADIOACTIVITY

Work Date: _____/_____/_____

LESSON OBJECTIVE

Students will explain how the "half-life" of atomic isotopes is used to find the age of ancient objects.

Classroom Activities

On Your Mark!

Draw Illustration C on the board and have students copy your diagram on Journal Sheet #3. Ask them to suppose that they have 16 kilograms of a mysterious "Element X." The substance is radioactive, giving off alpha and beta particles that you detect using a **Geiger counter**. Atom after atom loses an alpha or beta particle and the "click" of each particle as it passes through the Geiger counter is heard loud and clear. Students will read about the inventor of the Geiger counter, the German physicist **Hans Geiger** (b. 1882; d. 1945) on their Fact Sheet. Counting the clicks, we measure the rate at which the lump of element X loses radioactive particles. Let's say we detect 1 click per second on our counter. Using this information we can calculate the number of atoms of element X that

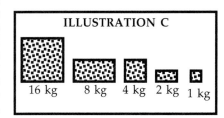

ILLUSTRATION C

16 kg   8 kg   4 kg   2 kg   1 kg

will disintegrate in a given amount of time. After doing a little bit of math, we determine that "half" of element X will be gone in one year at its present rate of disintegration. We say: "The **half-life** of element X is 1 year." Point out that neither intense heat nor cold will affect the rate of disintegration. So, in 4 years, there will be 1 gram of elements X left in the sample. Ask students to refer to the *Radioactive Half-Lives* section on their Fact Sheet and read the time it takes for half of each radioactive element on the list to decay.

Get Set!

Refer students to the activity described on Journal Sheet #3. Give them ample time to complete the activity.

Go!

After they have completed their graph propose the following situation. Have students imagine that the time between shakes of the box was 4 minutes. If a friend (i.e., like the school principal) entered the room at any time during the activity and counted the number of heads along with one group of students, could they tell us when the first shake took place? What information would our friend need to have? Answer: They would need to know the time between shakes (i.e., the rate at which half the pennies were lost) and the number of pennies we started with. With that information our friend could read backward on our graph and guess when the "shaking activity" began. If they counted 6 pennies left, they could conclude the following: 4 minutes ago there were 12 pennies; 8 minutes ago there were 24 pennies; 12 minutes ago there were 48 pennies and 16 minutes ago there were 96 (or approximately 100 pennies) which is the number of pennies we started with. Now suggest the following: If you have 1 gram of radioactive carbon 14 in you when you are alive and cease to recycle carbon 14 when you die, how old would your fossilized bones be when they contain 0.25 grams of carbon 14? Answer: 1 gram becomes 0.5 grams becomes 0.25 grams after 2 half-lives. If 1 half-life of carbon 14 is 5,730 years, your fossilized bones must be 11,460 years old! This is the method scientists use to estimate the age of ancient rocks and dinosaur bones.

Materials

100 pennies or M&M™s per group, shoe boxes

CH15 Journal Sheet #3

NUCLEAR REACTIONS AND RADIOACTIVITY

start   shake   shake   shake   shake   shake   shake
          #1   #2   #3   #4   #5   #6

RADIOACTIVE HALF LIFE

Directions: (1) Place 100 pennies or "M&M™s" "heads up" in a shoe box or similar container. (2) Plot the number "100" on your graph to indicate that you started with 100 pennies (or M&M™s). (3) Close and shake the box for ten seconds, then place it on the table and remove the pennies facing "tails up." (4) Plot the number of pennies left in the box on your graph, cover the box, and shake it again. (5) Continue the procedure until there is only one penny left.

CH15 Lesson #4
NUCLEAR REACTIONS AND RADIOACTIVITY

Work Date: ____/____/____

LESSON OBJECTIVE

Students will build models of atomic nuclei and illustrate the difference between fission and fusion.

Classroom Activities

On Your Mark!

If a supply of 100 dominoes is unavailable cut small blocks of wood with similar dimensions before the start of class.

Begin discussion by having students imagine the following scenario: A seemingly honest salesman offers them the following contract. If they will agree to invest 1 penny today, 2 pennies tomorrow, 4 pennies the following day and so on—doubling their investment every day for just 30 days—he will return them $5 million on their investment at the end of the month. Ask students if they would agree to sign such a contract; then have them number 1 through 30 down the margin of Journal Sheet #4 and do the calculation. As they perform their calculations make sure they are doubling the amount of money they need to give the salesman day after day, not simply adding 2 pennies per day. They will find that their total investment comes to considerably more than $5 million. They have been cheated! This mathematical geometric progression is called a **chain reaction** in **nuclear physics**.

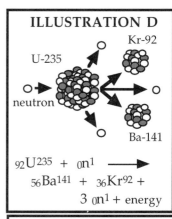

ILLUSTRATION D

Kr-92

U-235

neutron

Ba-141

$$_{92}U^{235} + {_0}n^1 \longrightarrow$$
$$_{56}Ba^{141} + {_{36}}Kr^{92} +$$
$$3 \, {_0}n^1 + energy$$

Get Set!

Draw Illustration D on the board and have students copy it on Journal Sheet #4. Point out that a neutron colliding with an unstable atom of uranium 235 splits the atom into barium 141 and krypton 92 in the nuclear reaction called a **fission reaction**. Energy and two more neutrons are expelled to split other atoms (i.e., $2 \times 2 \times 2 \times 2 \times 2$, etc.) producing a chain reaction that will result in the splitting of trillions of atoms in a matter of milli-seconds! Explain the nuclear equation at the bottom of the diagram. The great German-born American physicist **Albert Einstein** (b. 1879; d. 1955) determined that the amount of energy released by nuclear reactions can be calculated by the equation **$E=mc^2$**; where **E** is energy, **m** is the mass of the original substance, and **c** is the speed of light. Assist students in calculating the amount of energy that would be produced by 1 kilogram of U 235. Answer: E = $(1 \text{ kg}) \times (3 \times 10^8 \text{ m/s}) \times (3 \times 10^8 \text{ m/s}) = 9 \times 10^{16}$ joules of

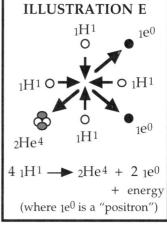

ILLUSTRATION E

$_1H^1$ $_1e^0$

$_1H^1$ $_1H^1$

$_2He^4$ $_1H^1$ $_1e^0$

$$4 \, {_1}H^1 \longrightarrow {_2}He^4 + 2 \, {_1}e^0$$
$$+ \; energy$$
(where $_1e^0$ is a "positron")

energy. That is enough energy to destroy a small city or supply it with electrical power for years. Atom bombs and modern nuclear reactors are based on the fission or "splitting" of atomic nuclei. Draw Illustration E on the board and have students copy it on Journal Sheet #4. Explain that the sun is powered by fusion. **Fusion** is the joining of light weight atomic nuclei, pressing them so close that the **strong force** glues them together. Great amounts of energy are released by this type of nuclear reaction as well according to Einstein's formula. The sun fuses more than 6 trillion tons of hydrogen into helium every second. It will use up all of its hydrogen in about 4–6 billion years.

Go!

Assist students in performing the chain reaction activity described in Figure A on Journal Sheet #4.

Materials

two colors of clay or M&M™s, 100 dominoes or small blocks of wood, stopwatches

Name: _____ Period: _____ Date: ___/___/___

CH15 JOURNAL SHEET #4
NUCLEAR REACTIONS AND RADIOACTIVITY

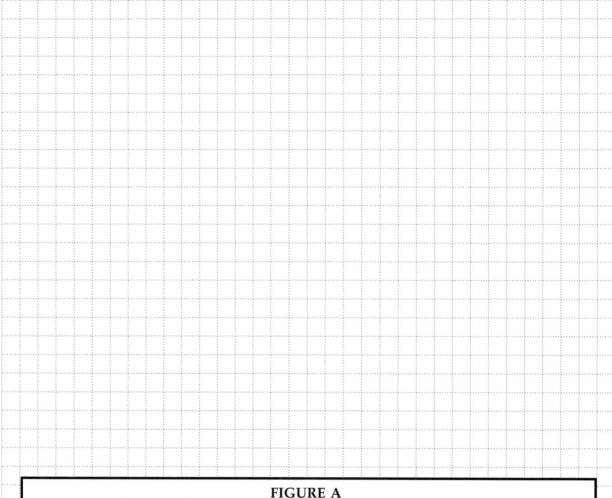

FIGURE A

Activity #1: (1) Use M&M™s or roll two different colors of clay into small clay "protons" and" neutrons." (2) Use the PERIODIC TABLE OF THE ELEMENTS to write the nuclear formulas for several small isotopes of your creation and bunch the "subatomic particles" into the "nuclei" of your chosen isotopes. (3) Show how fission and fusion reactions occur in your models to produce different atomic elements. (4) Write the equation for each nuclear reaction. Example: $_2He^4 + _7N^{14} \longrightarrow _1H^1 + _8O^{17}$.

Activity #2: (1) Place 50 dominoes or similarly shaped wooden blocks in a straight line so that tipping the first will cause the others to fall in sequence. (2) Arrange a second group of dominoes containing the same number of blocks or blocks as shown. (3) Set the stopwatch at zero and tip the first block in Group A. (4) Time how long it takes for every block to fall. (5) Reset the stopwatch and repeat the same action for Group B, timing the fall of all blocks as well. Which group fell faster? Why? How does this demonstration show the effects of a chain reaction?

TOP VIEW

Group A

Group B

NOTE: Make sure each group has the same number of blocks. It may take several trials before you are successful in causing all of the blocks to fall.

CH15 Review Quiz

Directions: Keep your eyes on your own work.
Read all directions and questions carefully.
THINK BEFORE YOU ANSWER!
Watch your spelling, be neat, and do the best you can.

CLASSWORK   (~40): _____
HOMEWORK   (~20): _____
CURRENT EVENT   (~10): _____
TEST   (~30): _____

TOTAL   (~100): _____
(A ≥ 90, B ≥ 80, C ≥ 70, D ≥ 60, F < 60)

LETTER GRADE: _____

TEACHER'S COMMENTS: _____
_____

NUCLEAR REACTIONS AND RADIOACTIVITY

TRUE–FALSE FILL-IN: If the statement is true, write the word TRUE. If the statement is false, change the underlined word to make the statement true. *24 points*

_____ 1. The nucleus of every atom except hydrogen contains protons and <u>electrons</u>.

_____ 2. Protons in the nucleus of an atom are <u>attracted to</u> one another.

_____ 3. Atomic nuclei are held together by a force called <u>gravity</u>.

_____ 4. Atoms <u>can</u> disintegrate.

_____ 5. Action of the <u>weak</u> force results in radioactivity.

_____ 6. <u>All</u> atoms with the same atomic number react in exactly the same way when mixed with other atoms.

_____ 7. Atoms with the same atomic number <u>always</u> have the same atomic mass.

_____ 8. Atoms of the same element that differ because of their mass are called <u>isotomers</u>.

_____ 9. <u>Beta</u> particles are negatively charged pieces of nuclear radiation.

_____10. An alpha particle has the same nuclear structure as a(n) <u>hydrogen</u> atom.

_____11. <u>Fission</u> is a process by which atomic nuclei are smashed and held together.

_____12. <u>Fusion</u> is a process by which atomic nuclei are broken apart.

PROBLEMS

NUCLEAR EQUATIONS: Refer to your PERIODIC TABLE OF THE ELEMENTS to show the missing product of each nuclear equation. *6 points*

13. p+ + beta particle → _____

14. $^{226}_{88}\text{Ra}$ → alpha particle + _____

15. $^{214}_{83}\text{Bi}$ → beta particle + _____

16. $^{14}_{6}\text{C}$ → beta particle + _____

17. $^{238}_{92}\text{U}$ → alpha particle + _____

18. $^{222}_{86}\text{Rn}$ → alpha particle + _____

_____   _____   ____/____/____
    Student's Signature   Parent's Signature   Date

APPENDIX

Keep this Grade Roster in the Science Section of your notebook

Date	Journal Points	Homework Points	Current Events Points	Quiz Points	Total Points	Letter Grade	Initials

HOW TO CALCULATE YOUR GRADE POINT AVERAGE

Your Report Card grades in this class will be awarded to you according to your grade point average or GPA. You can calculate your GPA whenever you like to find out exactly how you are doing in this class.

First, award each of your weekly grades the following credits: each A is worth 4 credits; each B is worth 3 credits; each C is worth 2 credits; each D is worth 1 credit; and each F is worth 0.

Add your total credits earned. Then, divide by the number of packets listed on your Grade Roster and round the decimal result to the nearest tenths place. Your overall Letter Grade is assigned according to the following GPA values:

A+ $\geq$ 4.0	A $\geq$ 3.7	A– $\geq$ 3.4
B+ $\geq$ 3.1	B $\geq$ 2.8	B– $\geq$ 2.5
C+ $\geq$ 2.2	C $\geq$ 1.9	C– $\geq$ 1.6
D+ $\geq$ 1.3	D $\geq$ 1.0	D– $\geq$ 0.7
	F $<$ 0.7	

FOR EXAMPLE:

John has completed five weeks of school and entered his grades from five packets on his Grade Roster. His grades are as follows: first week, A; second week, B; third week, C; fourth week, C; and fifth week, D.

John awards himself the correct amount of credit for each of his grades.

A	earns	4 credits
B	earns	3 credits
C	earns	2 credits
C	earns	2 credits
D	earns	1 credit
Total	earned is	12 credits

John divides his total credits earned by 5 (the number of packets on his Grade Roster).

12 divided by 5 equals 2.4

John's grade point average, or GPA, is 2.4. Referring to the grades shown above, John knows that he has a C+ in Science thus far, because 2.4 is greater than 2.2 (C+) but less than 2.5 (B–).

THE PERIODIC TABLE OF THE ELEMENTS

Legend

atomic number	6
chemical symbol	C
atomic mass	(12)

PHYSICAL PROPERTIES

Families 1-2 are light metals.
Families 3-7 are brittle.
Families 8-11 are ductile.
Family 12 is low boiling.
Families 13-17 are nonmetals.
Family 18 is inert.

Main table

	IA (1) alkali metals	IIA (2) alkaline metals	IIIA (3)	IVA (4)	VA (5)	VIA (6)	VIIA (7)	VIIIA (8)	(9)	(10)	IB (11)	IIB (12)	IIIB (13)	IVB (14)	VB (15)	VIB (16)	VIIB (17) halogens	VIIIB (18) noble gases
1	1 H (1)																	2 He (4)
2	3 Li (7)	4 Be (9)											5 B (11)	6 C (12)	7 N (14)	8 O (16)	9 F (19)	10 Ne (20)
3	11 Na (23)	12 Mg (24)											13 Al (27)	14 Si (28)	15 P (31)	16 S (32)	17 Cl (35)	18 Ar (40)
4	19 K (39)	20 Ca (40)	21 Sc (45)	22 Ti (48)	23 V (51)	24 Cr (52)	25 Mn (55)	26 Fe (56)	27 Co (59)	28 Ni (59)	29 Cu (63)	30 Zn (65)	31 Ga (70)	32 Ge (73)	33 As (75)	34 Se (79)	35 Br (80)	36 Kr (84)
5	37 Rb (85)	38 Sr (88)	39 Y (89)	40 91 (238)	41 Nb (93)	42 Mo (96)	43 Tc (97)	44 Ru (101)	45 Rh (103)	46 Pd (106)	47 Ag (108)	48 Cd (112)	49 In (114)	50 Sn (119)	51 Sb (122)	52 Te (128)	53 I (127)	54 Xe (131)
6	55 Cs (133)	56 Ba (137)	"L" series	72 Hf (178)	73 Ta (181)	74 W (184)	75 Re (186)	76 Os (190)	77 Ir (192)	78 Pt (195)	79 Au (197)	80 Hg (201)	81 Tl (204)	82 Pb (207)	83 Bi (209)	84 Po (210)	85 At (210)	86 Rn (222)
7	87 Fr (223)	88 Ra (226)	"A" series	104 Ku (251)	105 Ha (260)													

"L" or Lanthanide Series

57 La (139)	58 Ce (140)	59 Pr (141)	60 Nd (144)	61 Pm (145)	62 Sm (150)	63 Eu (152)	64 Gd (157)	65 Tb (159)	66 Dy (163)	67 Ho (165)	68 Er (167)	69 Tm (169)	70 Yb (173)	71 Lu (175)

"A" or Actinide Series

89 Ac (227)	90 Th (232)	91 Pa (231)	92 U (238)	93 Np (237)	94 Pu (242)	95 Am (243)	96 Cm (247)	97 Bk (249)	98 Cf (251)	99 Es (254)	100 Fm (257)	101 Md (256)	102 No (254)	103 Lr (257)

* The atomic number is equal to the number of protons in the nucleus of an atom.
* The atomic mass is equal to the total number of protons and neutrons in the nucleus of an element.
* The atomic mass is the mass in grams of 6×10^{23} atoms of an element.

Element names and symbols

actinium: Ac
aluminum: Al
americum: Am
antimony: Sb
argon: Ar
arsenic: As
astatine: At
barium: Ba
beryllium: Be
bismuth: Bi
boron: Bo
bromium: Br
cadmium: Cd
calcium: Ca
californium: Cf
carbon: C
cerium: Ce
cesium: Cs
chlorine: Cl
chromium: Cr
cobalt: Co
copper: Cu
curium: Cm
dysprosium: Dy
ensteinium: Es
erbium: Er
europium: Eu
fermium: Fm
fluorine: F
francium: Fr
gadolinium: Gd
gallium: Ga
germanium: Ge
gold: Au
hafnium: Hf
helium: He
holmium: Ho
hydrogen: H
indium: In
iodine: I
iridium: Ir
iron: Fe
krypton: Kr
kurchatovium: Ku
lanthanium: La
lawrencium: Lr
lead: Pb
lithium: Li
lutetium: Lu
magnesium: Mg
manganese: Mn
mendelevium: Md
mercury: Hg
molybdenum: Mo
neodymium: Nd
neon: Ne
neptunium: Np
nickel: Ni
niobium: Nb
nitrogen: N
nobelium: No
osmium: Os
oxygen: O
palladium: Pd
phosphorus: P
platinum: Pt
plutonium: Pu
polonium: Po
potassium: K
praseodymium: Pr
promethium: Pm
protactinium: Pa
radium: Ra
radon: Rn
rhenium: Re
rhodium: Rh
rubidium: Rb
ruthenium: Ru
samarium: Sm
scandium: Sc
selenium: Se
silicon: Si
silver: Ag
sodium: Na
strontium: Sr
sulfur: S
tantalium: Ta
technetium: Tc
tellurium: Te
terbium: Tb
thallium: Tl
thorium: Th
thulium: Tm
tin: Sn
titanium: Ti
tungsten: W
uranium: U
vanadium: V
xenon: Xe
ytterbium: Yb
Yttrium: Y
zinc: Zn
zirconium: Zr

Name: _____ **Period:** _____ **Date:** ____/____/____

Extra Journal Sheet

Fact Sheet Title: _____ **Lesson #** _____

LABORATORY SAFETY TECHNIQUES AND TIPS

The variety of techniques for preparing and handling equipment in the laboratory are as numerous as the scientists who practice them. Yet in all cases there is no substitute for common sense and diligent regard for health and safety. *Rehearse lab activities before presenting them to the class to better anticipate potential hazards.* Before beginning any laboratory demonstration or experiment be sure that students are aware of the following general safety guidelines:

- Be aware of the location and proper use of safety equipment such as the classroom fire extinguisher, dust pan and broom, eye wash or shower.

- Wear goggles and protective clothing (e.g., gloves, apron) during near-at-hand demonstrations and all experiments.

- Clean up after experiments by washing and drying hands, counter tops and labware.

- Ask for help if you are not exactly sure how to conduct any part of an experiment.

- Tell the teacher immediately if an accident occurs.

Handling Chemicals

Assume that all reagents are toxic and dangerous. Avoid direct contact with all chemicals by using protective goggles, aprons, and gloves.

Never attempt to identify a liquid by smelling it. Even mild reagents can be irritating to the eyes and linings of the nose and mouth. To transfer **liquids** from a labelled container, turn the container so that the label faces up toward your palm. As you pour, position your thumb and fingers so that they are not in the path of the liquid as it drips down the underside of the container. Some scientists choose to insert a glass rod into the receiving container and pour the liquid down the rod.

Avoid direct contact with **solid** reagents. Use spatulas (i.e., wooden tongue depressors or straws cut lengthwise) to transfer solids. Fold a piece of wax paper and place it on a balance pan to weigh solids. Transfer the solid inside the folded wax paper to the receiving container. Or, weigh the container, then add the appropriate amount of solid to the container while it is still on the balance.

When experimenting with **gases**, be sure that the room is well ventilated. If a ventilation hood is not available, wear a surgical mask.

Cutting Glass Tubing

Using a triangular file, "saw" a neat line at a right angle to the length of the tube. Grasp the tube with both hands, thumbs together at the file line so that the line is facing away from you. Apply gentle pressure away from you with both thumbs. Hold the tips of the glass over a Bunsen burner to smooth the edges of the fractured glass.

Bending Glass Tubing

Hold the length of tubing to be bent with both hands over a "wing topped" Bunsen burner. Be sure there is at least 8 to 12 inches between your hands and the flame. Roll the tube between the thumbs, index, and middle fingers to heat the tube completely around. When the glass begins to sag remove it from the flame and hold it with one hand horizontal to the floor allowing gravity to bend the tube for you. Place the bent glass on a heat-resistant surface and allow it to cool.

Inserting Glass Tubing into a Rubber Stopper

Dip the end of the glass tube in glycerine or vaseline. Hold it within an inch of the stopper hole. Gently "screw" the glass into the stopper. Wipe the glass clean.

Connecting Rubber Tubing to Glass Tubing

Grasp the glass tubing between thumb and forefinger no more than one centimeter from the end. Gently twist the rubber tubing onto the end of the glass.

Capping a Test Tube with a Rubber Stopper

Secure the test tube in a test tube rack or in a clamp secured to a ring stand. Grasp the test tube with thumb and forefinger no more than one centimeter from the end. Gently twist the stopper into the tube until it is snug. It is not necessary to press the stopper down into the test tube. This will make it difficult if not impossible to remove.

Preparing Acidic Solutions

ALWAYS pour acids into water or other aqueous solutions. NEVER pour water or aqueous solution into acids! Pouring acid into water dilutes the acid as it goes into solution. Pouring water into acid causes the hydrogen ions in the acid to compete vigorously for the water molecules thereby producing a heated reaction. Use the technique described above for handling liquids.

Housekeeping

In most cases warm soapy water is adequate for cleaning counter tops and glassware. Stained glassware can be soaked in a mild sulfuric acid solution (i.e., 10%) for several days. If available, allow the glassware to soak under a ventilation hood. Water will form a continuous film on clean glassware and spot on dirty glassware.

How to Light a Bunsen Burner

Wear goggles and clothing that fits tightly to the body. Sagging sleeves can present a fire hazard. Tie back long hair. Turn off the main gas valve on the counter or gas tank. Be sure that the Bunsen burner tube is clear of obstructions (e.g., wadded paper, dirt). Set the Bunsen tube about 5 mm above the gas inlet hole. Open the Bunsen gas valve "one twist" from the closed position. Stay clear of the burner. Turn on the main gas valve and use a match or sparker to light the flame at the top of the Bunsen burner tube. Use the Bunsen gas valve to control the amount of gas reaching the end of the tube. Adjust the tube height to control the addition of atmospheric oxygen to the flame until the flame has a clear blue hue. Too little oxygen results in a wavering yellow flame. Use the main gas valve to shut off an "unruly" flame. NEVER TURN YOUR BACK TO A LIT BURNER!

USING CURRENT EVENTS TO INTEGRATE SCIENCE INSTRUCTION ACROSS CONTENT AREAS

Science does not take place in a vacuum. Scientists, like other professionals, are influenced by the economic and political realities of their time. In addition, the ideological and technological advances made by science can influence the economic and political structure of society—for better or worse. It is therefore essential that students are aware of the day-to-day science being conducted at laboratories around the world—important work being reported by an international news media.

Most State Departments of Education, make **CURRENT EVENTS** a regular part of their state science frameworks. Science instructors can use newspaper, magazine, and television reports to keep their students informed about the advances and controversies stemming from research in the many scientific disciplines. Teachers can also use current events to integrate science instruction across the curriculum.

Set aside a class period to show students how to prepare a **science** or **technology** current event. They can do this on a single sheet of standard looseleaf paper. You may require pupils to read all or part of a science/technology article depending on their reading level. Have them practice summarizing the lead and one or more paragraphs of the article *in their own words*. Advise them to keep a **thesaurus** on hand or to use the dictionary/thesaurus stored in their personal computer at home. Tell students to find *synonyms* they can use to replace most of the vocabulary words used by the article's author. This activity will help them to expand their vocabulary and improve grammar skills. Show students how to properly trim and paste the article's title and first few paragraphs on the front of a standard piece of looseleaf. They should write their summary on the opposite side of the page so that the article is visible to their classmates when they make an oral presentation to the class. Allow students to make a report that summarizes a newsworthy item they may have heard on television. The latter report should be accompanied by the signature of a parent/guardian to insure the accuracy of the information being presented.

Students' skills at public speaking are sure to improve if they are given an opportunity to share their current event. Current events can be shared after the end-of-the-unit REVIEW QUIZ or whenever the clock permits at the end of a lesson that has been completed in a timely fashion. You can select students at random to make their presentations by drawing lots or ask for volunteers who might be especially excited about their article. Take time to discuss the ramifications of the article and avoid the temptation to express your personal views or bias. Remain objective and give students the opportunity to express their views and opinions. Encourage them to base their views on fact: not superstition or prejudice. Should the presentation turn into a debate, set aside a few minutes later in the week, giving students time to prepare what they would like to say. Model courtesy and respect for all points of view and emphasize the proper use of the English language in all modes of presentation: both written and oral.

BIO-DATA
CARDS

INSTRUCTIONS TO TEACHERS
Xerox and cut out the Bio-Data Cards below and keep them in a handy file. Instruct students to choose one card and neatly glue it to the front of a 5" × 8" index card. They can use the school or public library to find out more about the scientist they have chosen. On the back of the index card they can draw a cartoon, write a poem or short paragraph that illustrates an important event in the life of this famous personality.

BIO-DATA CARD
ARISTOTLE
(born 384 B.C.; died 322 B.C.)

nationality
Greek

contribution to science
advocated reason and maintained
that all we know of the universe
comes from our senses

BIO-DATA CARD
SVANTE AUGUST ARRHENIUS
(born 1859; died 1927)

nationality
Swiss

contribution to science
founded physical chemistry and
predicted global warming from the
burning of fossil fuels

BIO-DATA CARD
AMADEO AVOGADRO
(born 1776; died 1856)

nationality
Italian

contribution to science
showed scientists how to calculate the
number of atoms or molecules
in a given amount of gas

BIO-DATA CARD
LEO HENDRICK BAEKELAND
(born 1863; died 1944)

nationality
Belgian-born American

contribution to science
invented the first commercial
plastic called Bakelite from
formaldehyde and phenol

BIO-DATA CARD
HENRI BECQUEREL
(born 1852; died 1908)

nationality
French

contribution to science
discovered radioactivity
in uranium ores

BIO-DATA CARD
CLAUDE BERNARD
(born 1813; died 1878)

nationality
French

contribution to science
founder of experimental medicine for
discovering the basic functions of the
intestine and the liver

BIO-DATA CARD
JÖNS JAKOB BERZELIUS
(born 1779; died 1848)

nationality
Swedish

contribution to science
used the term "organic" to describe
chemicals in living things and created
the system for writing chemical symbols

BIO-DATA CARD
JOSEPH BLACK
(born 1728; died 1799)

nationality
Scottish

contribution to science
discovered carbon dioxide and laid
the foundation for the work of James Watt
who improved the steam engine

INSTRUCTIONS TO TEACHERS
Xerox and cut out the Bio-Data Cards below and keep them in a handy file. Instruct students to choose one card and neatly glue it to the front of a 5″ × 8″ index card. They can use the school or public library to find out more about the scientist they have chosen. On the back of the index card they can draw a cartoon, write a poem or short paragraph that illustrates an important event in the life of this famous personality.

BIO-DATA CARD

NIELS BOHR
(born 1885; died 1962)

nationality
Danish

contribution to science
elucidated the structure of the atom
and became the father of
quantum mechanics

BIO-DATA CARD

LUDWIG EDUARD BOLTZMAN
(born 1844; died 1906)

nationality
Austrian

contribution to science
invented statistical mechanics
to describe the motion of atoms and
molecules in gases

BIO-DATA CARD

JEAN-BAPTISTE BOUSSINGAULT
(born 1802; died 1887)

nationality
French

contribution to science
showed that plants obtained nitrogen
from the soil, not the air

BIO-DATA CARD

ROBERT BOYLE
(born 1627; died 1691)

nationality
Irish

contribution to science
discovered Boyle's Law which predicts
the volume or pressure of gases at a
particular temperature

BIO-DATA CARD

WILLIAM & LAWRENCE BRAGG
(born 1862; died 1942) (born 1890; died 1971)

nationality
Australian-English

contribution to science
father and son pioneers in the field of
X-ray crystallography

BIO-DATA CARD

ROBERT BROWN
(born 1773; died 1858)

nationality
Scottish

contribution to science
discovered the random motion of
microscopic particles later described as
Brownian motion

BIO-DATA CARD

ROBERT WILLIAM BUNSEN
(born 1811; died 1899)

nationality
German

contribution to science
invented the Bunsen burner and
discovered the elements
cesium and rubidium

BIO-DATA CARD

WALLACE HUME CAROTHERS
(born 1896; died 1937)

nationality
American

contribution to science
the first to polymerize nylon fibers
and polyesters that replaced silk
as a material for clothing

BIO-DATA CARD

HENRY CAVENDISH
(born 1731; died 1810)

nationality
English

contribution to science
discovered hydrogen
and determined the composition
of water and nitric acid

BIO-DATA CARD

ANDERS CELSIUS
(born 1701; died 1744)

nationality
Swedish

contribution to science
invented the Celsius/Centigrade scale to measure temperature where water freezes at 0° and boils at 100°

BIO-DATA CARD

JAMES CHADWICK
(born 1891; died 1974)

nationality
English

contribution to science
discovered the neutral particles in the nuclei of atoms later called neutrons

BIO-DATA CARD

HILAIRE B. CHARDONNET
(born 1839; died 1924)

nationality
French

contribution to science
produced the first artificial fiber
called rayon

BIO-DATA CARD

MICHEL-EUGÉNE CHEVRUEL
(born 1786; died 1889)

nationality
French

contribution to science
developed a procedure for assessing pure substances and discovered margaric acid used to make margarine

BIO-DATA CARD

JOHN DOUGLAS COCKCROFT
(born 1897; died 1967)

nationality
English

contribution to science
first to succeed in splitting an atom with his associate Ernest Walton

BIO-DATA CARD

FRANCIS H. C. CRICK
(born 1916)

nationality
English

contribution to science
working with James Watson discovered the structure of the DNA molecule

BIO-DATA CARD

MARIE CURIE
(a.k.a. Manya Sklodowska)
(born 1867; died 1934)

nationality
Polish-French

contribution to science
discovered the radioactive elements
polonium and radium

BIO-DATA CARD

PIERRE CURIE
(born 1859; died 1906)

nationality
French

contribution to science
shared the Nobel Prize in 1903 with his wife Marie Curie for the discovery of polonium and radium

BIO-DATA CARD

LOUIS DAGUERRE
(born 1769; died 1851)

nationality
French

contribution to science
invented the first reliable method for producing photographic images on copper plates

BIO-DATA CARD

JOHN DALTON
(born 1766; died 1844)

nationality
English

contribution to science
proposed the atomic theory which described atoms as the fundamental particles of matter

BIO-DATA CARD

ABRAHAM DARBY
(born 1677; died 1717)

nationality
English

contribution to science
perfected the first efficient method for smelting iron with coke instead of charcoal

BIO-DATA CARD

CHARLES ROBERT DARWIN
(born 1809; died 1882)

nationality
English

contribution to science
developed the theory of evolution

BIO-DATA CARD

HUMPHRY DAVY
(born 1778; died 1829)

nationality
English

contribution to science
used electrolysis to discover the elements boron, calcium, magnesium, potassium, and sodium

BIO-DATA CARD

DEMOCRITUS
(born 460 B.C.; died 370 B.C.)

nationality
Greek

contribution to science
introduced the idea that matter derived from tiny indivisible particles

BIO-DATA CARD

RENÉ DESCARTES
(born 1596; died 1650)

nationality
French

contribution to science
one of the first respected philosophers to propose that physiological functions could be explained by physical laws

BIO-DATA CARD

EDWIN LAURENTINE DRAKE
(born 1819; died 1880)

nationality
American

contribution to science
drilled the first oil well in Titusville, USA, in 1859

BIO-DATA CARD

PAUL EHRLICH
(born 1854; died 1915)

nationality
German

contribution to science
founder of chemotherapy

BIO-DATA CARD

ALBERT EINSTEIN
(born 1879; died 1955)

nationality
German-American

contribution to science
determined the formula $E=mc^2$ to calculate the amount of energy contained in the nuclei of atoms

BIO-DATA CARD

EMPEDOCLES
(born 493 BC; died 433 B.C.)

nationality
Greek

contribution to science
proposed that the universe was composed of four basic elements: air, earth, fire, and water

BIO-DATA CARD

GABRIEL DANIEL FAHRENHEIT
(born 1686; died 1736)

nationality
Polish-Dutch

contribution to science
invented the first accurate thermometer in 1724

BIO-DATA CARD

MICHAEL FARADAY
(born 1791; died 1867)

nationality
English

contribution to science
isolated benzene and showed that platinum was useful as a catalyst in speeding up chemical reactions

BIO-DATA CARD

ENRICO FERMI
(born 1901; died 1954)

nationality
Italian-American

contribution to science
built the first atomic fission reactor in Chicago in 1942

BIO-DATA CARD

ELIZABETH FULHAME
(published in 1794)

nationality
English or Irish

contribution to science
succeeded in making the first photographic images on cloth using gold and silver salts

BIO-DATA CARD

GALEN
(born 129; died 200)

nationality
Greek

contribution to science
developed a medical theory based on imaginary body humours that guided the field of medicine for 1,500 years

BIO-DATA CARD

GALILEO GALILEI
(born 1564; died 1642)

nationality
Italian

contribution to science
invented the first thermoscope to measure the temperature of the air

BIO-DATA CARD

PIERRE GASSENDI
(born 1592; died 1655)

nationality
French

contribution to science
promoted the theory of atomism

BIO-DATA CARD

HANS GEIGER
(born 1882; died 1945)

nationality
German

contribution to science
invented the Geiger counter for detecting radioactive particle emissions from the nuclei of atoms

BIO-DATA CARD

DONALD ARTHUR GLASER
(born 1926)

nationality
American

contribution to science
invented the bubble chamber for detecting high energy elementary particles

BIO-DATA CARD

JOHANN RUDOLF GLAUBER
(born 1604; died 1670)

nationality
German

contribution to science
first to manufacture pure, concentrated hydrochloric acid to be used in the preparation of commercial salts

BIO-DATA CARD

WILLIAM HARVEY
(born 1578; died 1657)

nationality
English

contribution to science
demonstrated how the heart circulates blood throughout the body through a system of arteries and veins

BIO-DATA CARD

RENÉ-JUST HAÜY
(born 1743; died 1822)

nationality
French

contribution to science
founded modern crystallography by proposing that molecules assembled themselves in geometric patterns

BIO-DATA CARD

WERNER KARL HEISENBERG

(born 1901; died 1976)

nationality
German

contribution to science
developed quantum theory and
determined the probability
of positioning particles within an atom

BIO-DATA CARD

GERMAIN HESS

(born 1802; died 1850)

nationality
Russian

contribution to science
measured heat exchanges during
chemical reactions to become a founder
of the field of thermochemistry

BIO-DATA CARD

HIPPOCRATES

(born 460 B.C.; died 377 B.C.)

nationality
Greek

contribution to science
called the founder and father
of medicine for his ideas
on cleanliness and diet

BIO-DATA CARD

ROBERT HOOKE

(born 1635; died 1703)

nationality
English

contribution to science
published the first thorough description
of minerals with crystalline structure

BIO-DATA CARD

ERNST F. HOPPE-SEYLER

(born 1825; died 1895)

nationality
German

contribution to science
isolated hemoglobin and founded the
first journal of biochemistry

BIO-DATA CARD

JOHN WESLEY HYATT

(born 1837; died 1920)

nationality
American

contribution to science
invented celluloid, a strong thin film of
plastic later used in the making of
motion picture film

BIO-DATA CARD

JAN INGENHOUSZ

(born 1730; died 1799)

nationality
Dutch

contribution to science
determined that plants absorb
carbon dioxide and give off oxygen
when exposed to sunlight

BIO-DATA CARD

WILLIAM THOMSON KELVIN

(born 1824; died 1907)

nationality
Irish

contribution to science
introduced an absolute temperature scale
by determining the lowest limit of
molecular motion

BIO-DATA CARD

IRVING LANGMUIR
(born 1881; died 1957)

nationality
American

contribution to science
the first to provide
an accurate measure of the size
of molecules

BIO-DATA CARD

PIERRE SIMON LAPLACE
(born 1749; died 1827)

nationality
French

contribution to science
helped Antoine Laurent Lavoisier show
that water was a compound and not an
element as was thought by the Greeks

BIO-DATA CARD

AUGUST LAURENT
(born 1807; died 1853)

nationality
French

contribution to science
discovered important organic compounds

BIO-DATA CARD

ANTOINE LAURENT LAVOISIER
(born 1743; died 1794)

nationality
French

contribution to science
proved that oxygen, which he named,
was necessary for combustion and
showed that water was a compound

BIO-DATA CARD

ANTON von LEEUWENHOEK
(born 1632; died 1732)

nationality
Dutch

contribution to science
invented the microscope

BIO-DATA CARD

GOTTFRIED WILHELM LEIBNIZ
(born 1646; died 1716)

nationality
German

contribution to science
designed the first calculating machine

BIO-DATA CARD

JUSTUS von LIEBIG
(born 1803; died 1873)

nationality
German

contribution to science
showed that plants absorb
minerals from the soil and mixed the
first artifical fertilizer

BIO-DATA CARD

CAROLUS LINNAEUS
(born 1707; died 1788)

nationality
Swedish

contribution to science
introduced the binomial classification
system to categorize groups of living
things

INSTRUCTIONS TO TEACHERS
Xerox and cut out the Bio-Data Cards below and keep them in a handy file. Instruct students to choose one card and neatly glue it to the front of a 5″ × 8″ index card. They can use the school or public library to find out more about the scientist they have chosen. On the back of the index card they can draw a cartoon, write a poem or short paragraph that illustrates an important event in the life of this famous personality.

BIO-DATA CARD

ANDREAS S. MARGGRAFF

(born 1709; died 1782)

nationality
Dutch

contribution to science
set down rules for determining the purity of substances and considered "the father of the sugar industry"

BIO-DATA CARD

JAMES CLERK MAXWELL

(born 1831; died 1879)

nationality
Scottish

contribution to science
proved that heat determined the motion of atoms and molecules making up matter

BIO-DATA CARD

ELMER VERNER McCOLLUM

(born 1879; died 1967)

nationality
American

contribution to science
determined and named important vitamins and minerals essential to a nutritional diet

BIO-DATA CARD

FERDINANDO DÉ MEDICI

(born 1610; died 1670)

nationality
Italian

contribution to science
invented the first sealed thermometer

BIO-DATA CARD

DMITRI I. MENDELEEV

(born 1834; died 1907)

nationality
Russian

contribution to science
developed the periodic law of chemistry and created the first authoritative chart of the elements

BIO-DATA CARD

LOTHAR MEYER

(born 1830; died 1895)

nationality
German

contribution to science
developed a periodic law of the elements independent of Dmitri Mendeleev

BIO-DATA CARD

STANLEY LLOYD MILLER

(born 1930)

nationality
American

contribution to science
with Harold Clayton Urey showed that amino acids and proteins could be made from simpler chemical compounds

BIO-DATA CARD

HENRY G. J. MOSELEY

(born 1887; died 1915)

nationality
English

contribution to science
revised THE PERIODIC TABLE according to increasing atomic number (i.e., the ratio of positive charges inside atoms)

BIO-DATA CARD

JOHN A. R. NEWLANDS
(born 1837; died 1898)

nationality
English

contribution to science
first to suggest that elements behaved in a "periodic" manner according to their increasing atomic masses

BIO-DATA CARD

SIR ISAAC NEWTON
(born 1642; died 1727)

nationality
English

contribution to science
popularized the notion that objects were attracted and repelled by invisible forces

BIO-DATA CARD

J. ROBERT OPPENHEIMER
(born 1904; died 1967)

nationality
American

contribution to science
coordinated the construction of the first atomic bomb in Los Alamos, New Mexico, 1945

BIO-DATA CARD

ALEXANDER PARKES
(born 1813; died 1890)

nationality
English

contribution to science
invented the first plastic from the plant extract cellulose nitrate

BIO-DATA CARD

LOUIS PASTEUR
(born 1822; died 1895)

nationality
French

contribution to science
discovered a variety of germs and popularized the notion that disease was caused by microorganisms

BIO-DATA CARD

LINUS CARL PAULING
(born 1901; died 1994)

nationality
American

contribution to science
showed how chemical bonds determined the shapes of large biochemical molecules

BIO-DATA CARD

MAX KARL ERNST PLANCK
(born 1858; died 1947)

nationality
German

contribution to science
laid the basis of quantum theory from his work on radiating black bodies

BIO-DATA CARD

JOSEPH PRIESTLEY
(born 1733; died 1804)

nationality
English

contribution to science
identified oxygen in 1774 as well as other gases and produced a method for making sulphur dioxide

INSTRUCTIONS TO TEACHERS
Xerox and cut out the Bio-Data Cards below and keep them in a handy file. Instruct students to choose one card and neatly glue it to the front of a 5″ × 8″ index card. They can use the school or public library to find out more about the scientist they have chosen. On the back of the index card they can draw a cartoon, write a poem or short paragraph that illustrates an important event in the life of this famous personality.

BIO-DATA CARD

JOSEPH LOUIS PROUST

(born 1754; died 1826)

nationality
French

contribution to science
proposed and developed the idea that compounds were composed of elements in specific ratios

BIO-DATA CARD

WILLIAM PROUT

(born 1785; died 1850)

nationality
English

contribution to science
proposed that the mass of every atom was a multiple of the mass of hydrogen and for whom the proton is named

BIO-DATA CARD

THEODORE W. RICHARDS

(born 1868; died 1928)

nationality
American

contribution to science
introduced a number of analytical techniques for determining the masses of atoms

BIO-DATA CARD

GILLES P. de ROBERVAL

(born 1602; died 1675)

nationality
French

contribution to science
invented a balance that allowed the accurate measuring of samples

BIO-DATA CARD

WILHELM KONRAD RÖNTGEN

(born 1845; died 1923)

nationality
German

contribution to science
discovered X-rays in 1895 which revolutionized the field of medical diagnosis

BIO-DATA CARD

BENJAMIN T. RUMFORD

(born 1753; died 1814)

nationality
English

contribution to science
proved that heat was not a substance but a product of friction and a form of energy

BIO-DATA CARD

ERNST RUTHERFORD

(born 1871; died 1937)

nationality
New Zealand-British

contribution to science
clarified the nuclear structure of the atom, discovered the proton and alpha and beta particles

BIO-DATA CARD

JULIUS von SACHS

(born 1832; died 1897)

nationality
German

contribution to science
showed that photosynthesis occurs in the chloroplasts of plant cells

BIO-DATA CARD

HORACE & NICHOLAS de SAUSSURE
(born 1740; died 1799) (born 1767; died 1845)

nationality
Swiss

contribution to science
father and son who clarified the nitrogen and carbon-oxygen cycles in plants

BIO-DATA CARD

ANDREAS F. W. SCHIMPER
(born 1856; died 1901)

nationality
German

contribution to science
showed that plants stored energy from the sun in starch molecules

BIO-DATA CARD

MATTHIAS JAKOB SCHLEIDEN
(born 1804; died 1881)

nationality
German

contribution to science
discovered that all living matter is made up of cells

BIO-DATA CARD

JEAN SENEBIER
(born 1742; died 1809)

nationality
Swiss

contribution to science
demonstrated that the cycling of carbon dioxide and oxygen in plants was a "light dependent" process

BIO-DATA CARD

SØREN P. L. SØRENSEN
(born 1868; died 1939)

nationality
Danish

contribution to science
introduced the idea of using hydrogen ion concentration to measure acidity and alkalinity (the pH scale)

BIO-DATA CARD

GEORG ERNST STAHL
(born 1660; died 1734)

nationality
German

contribution to science
developed the false but productive theory that substances burned because they contained flammable "phlogiston"

BIO-DATA CARD

JEAN SERVAIS STAS
(born 1813; died 1891)

nationality
Belgian

contribution to science
made the first accurate measurements of atomic masses according to the elements' relative weights

BIO-DATA CARD

KEKÚLE von STRADONITZ
(born 1829; died 1896)

nationality
German

contribution to science
revolutionized organic chemistry with his determination of the structures of methane and benzene

---

BIO-DATA CARD

THALES
(born 624 B.C.; died 547 B.C.)

nationality
Greek

contribution to science
first pre-Socratic philosopher to suggest
that water was recycled
throughout the environment

BIO-DATA CARD

JOSEPH JOHN THOMSON
(born 1856; died 1940)

nationality
English

contribution to science
discovered the electron in 1897
and laid the basis for an
electrical theory of the atom

BIO-DATA CARD

HAROLD CLAYTON UREY
(born 1893; died 1981)

nationality
American

contribution to science
discovered deuterium, an isotope of
hydrogen, and developed theories about
the formation of the Earth

BIO-DATA CARD

ANDREAS VESALIUS
(born 1514; died 1564)

nationality
Belgian

contribution to science
made the first rigorous studies of human
anatomy

BIO-DATA CARD

JOHANNES van der WAALS
(born 1837; died 1923)

nationality
Dutch

contribution to science
explained how electrostatic forces hold
molecules together

BIO-DATA CARD

ERNEST WALTON
(born 1903; died 1995)

nationality
Irish

contribution to science
collaborated with John Cockcroft
to achieve the first successful
splitting of atoms

BIO-DATA CARD

JAMES DEWEY WATSON
(born 1928)

nationality
American

contribution to science
collaborated with Francis Crick
to determine the structure
of the DNA molecule

BIO-DATA CARD

JAMES WATT
(born 1736; died 1819)

nationality
Scottish

contribution to science
improved the steam engine
which led to the understanding of heat as
a form of kinetic energy

BIO-DATA CARD

HEINRICH OTTO WIELAND
(born 1877; died 1957)

nationality
German

contribution to science
determined the structure of organic steroids and showed that bile is a form of cholesterol

BIO-DATA CARD

RICHARD WILLSTÄTTER
(born 1872; died 1942)

nationality
German

contribution to science
determined the structure of chlorophyll, laying the foundation of the pharmaceutical industry

BIO-DATA CARD

CHARLES T. R. WILSON
(born 1869; died 1959)

nationality
Scottish

contribution to science
invented the Wilson cloud chamber to track the motion of subatomic particles

BIO-DATA CARD

FRIEDRICH WÖHLER
(born 1800; died 1882)

nationality
German

contribution to science
the first to synthesize urea to show that molecules present in living things could be derived from nonliving matter

BIO-DATA CARD

WILLIAM HYDE WOLLASTON
(born 1766; died 1828)

nationality
English

contribution to science
suggested that understanding the three dimensional structure of atoms would clarify how chemical reactions occurred